EUROPT – A European Initiative on Optimum Design Methods in Aerodynamics

Edited by
Jacques Periaux
Gabriel Bugeda
Panagiotis K. Chaviaropoulos
Theo Labrujere
and Bruno Stoufflet

Notes on Numerical Fluid Mechanics (NNFM) Volume 55

Volume 54 Boundary Elements: Implementation and Analysis of Advanced Algorithms. Proceesings of the Twelfth GAMM-Seminar, Kiel, January 19–21, 1996 (W. Hackbusch / G. Wittum, Eds.)
Volume 53 Computation of Three-Dimensional Complex Flows. Proceedings of the IMACS-COST Conference on Computational Fluid Dynamics, Lausanne, September 13–15, 1995 (M. Deville / S. Gavrilakis / I. L. Ryhming, Eds.)
Volume 52 Flow Simulation with High-Performance Computers II. DFG Priority Research Programme Results 1993–1995 (E. H. Hirschel, Ed.)
Volume 51 Numerical Treatment of Coupled Systems. Proceedings of the Eleventh GAMM-Seminar, Kiel, January 20–22, 1995 (W. Hackbusch / G. Wittum, Eds.)
Volume 50 Computational Fluid Dynamics on Parallel Systems. Proceedings of a CNRS-DFG Symposium in Stuttgart, December 9 and 10, 1993 (S. Wagner, Ed.)
Volume 49 Fast Solvers for Flow Problems. Proceedings of the Tenth GAMM-Seminar, Kiel, January 14–16, 1994 (W. Hackbusch / G. Wittum, Eds.)
Volume 48 Numerical Simulation in Science and Engineering. Proceedings of the FORTWIHR Symposium on High Performance Scientific Computing, München, June 17–18, 1993 (M. Griebel / Ch. Zenger, Eds.)
Volume 47 Numerical Methods for the Navier-Stokes Equations (F.-K. Hebeker, R. Rannacher, G. Wittum, Eds.)
Volume 46 Adaptive Methods – Algorithms, Theory and Applications. Proceedings of the Ninth GAMM-Seminar, Kiel, January 22–24, 1993 (W. Hackbusch / G. Wittum, Eds.)
Volume 45 Numerical Methods for Advection – Diffusion Problems (C. B. Vreugdenhil / B. Koren, Eds.)
Volume 44 Multiblock Grid Generation – Results of the EC/BRITE-EURAM Project EUROMESH, 1990–1992 (N. P. Weatherill / M. J. Marchant / D. A. King, Eds.)
Volume 43 Nonlinear Hyperbolic Problems: Theoretical, Applied, and Computational Aspects Proceedings of the Fourth International Conference on Hyperbolic Problems, Taormina, Italy, April 3 to 8, 1992 (A. Donato / F. Oliveri, Eds.)
Volume 42 EUROVAL – A European Initiative on Validation of CFD Codes (W. Haase / F. Brandsma / E. Elsholz / M. Leschziner / D. Schwamborn, Eds.)
Volume 41 Incomplete Decompositions (ILU) – Algorithms, Theory and Applications (W. Hackbusch / G. Wittum, Eds.)
Volume 40 Physics of Separated Flow – Numerical, Experimental, and Theoretical Aspects (K. Gersten, Ed.)
Volume 39 3-D Computation of Incompressible Internal Flows (G. Sottas / I. L. Ryhming, Eds.)
Volume 38 Flow Simulation on High-Performance Computers I (E. H. Hirschel, Ed.)
Volume 37 Supercomputers and their Performance in Computational Fluid Mechanics (K. Fujii, Ed.)
Volume 36 Numerical Simulation of 3-D Incompressible Unsteady Viscous Laminar Flows (M. Deville / T.-H. Lê / Y. Morchoisne, Eds.)

Volumes 1 to 29, 45 are out of print.
The addresses of the Editors and further titles of the series are listed at the end of the book.

EUROPT – A European Initiative on Optimum Design Methods in Aerodynamics

Proceedings of the Brite/Euram Project Workshop "Optimum Design in Aerodynamics", Barcelona, 1992

Edited by
Jacques Periaux
Gabriel Bugeda
Panagiotis K. Chaviaropoulos
Theo Labrujere
and Bruno Stoufflet

Softcover reprint of the hardcover 1st edition 1997

Vieweg is a subsidiary company of the Bertelsmann Professional Information.

Produced by W. Langelüddecke, Braunschweig
Printed on acid-free paper

ISSN 0179-9614
ISBN-13: 978-3-322-86572-4 e-ISBN-13: 978-3-322-86570-0
DOI: 10.1007/978-3-322-86570-0

PREFACE

This volume entitled "EUROPT - A European Initiative on Optimum Design in Aerodynamics" contains the results of the contributors during a workshop which took place in Barcelona in June 1992. This workshop was organized in the framework of the Brite/Euram Aeronautics project "Optimum Design in Aerodynamics" (AERO-89-0026). The project brought together nine European partners from Academy and Industry with big experience in numerical optimization techniques applied to automated optimum design.

The manuscript is directed to the optimization field and its goal is to provide the reader with useful numerical optimization techniques for optimum design in aerodynamics.

The field of numerical optimization techniques has been growing since the early 50's. But only recently, with the recent advent of powerful computers, it has yielded real-life applications that demonstrate industrial potential for design purposes.

Nowadays, optimization is at the stage where those scientists and engineers who have worked in the field are to be paid to apply numerical optimization to automated design of real problems, namely aerodynamic shapes in Aerospace Engineering.

For the above reasons, it was evident that a Workshop on numerical optimization techniques for the validation of optimum design methods would be of interest to most of the partners involved in the AERO-89-0026 project in order to compare in terms of accuracy and efficiency several optimization softwares performed on the same selected flow problems.

A list of test cases representing the selected flow problems was defined. These test cases can be grouped in the three following areas: reconstruction test cases, inverse test cases and optimization problems - namely reduction of wave drag for nozzles, airfoils or wings - evolving in inviscid subsonic or transonic flows.

Since optimum design requires very robust methodologies for a constrained space of research, the workshop was aimed to give in depth analysis of different numerical optimization techniques used for different classes of design ranging from academic test cases to industrial ones by their dimensionality or their complexity (some of them belonging for example to the family of transonic multi point design).

The participants in the workshop prepared a document based on the results of their presented test cases computations with the description of the methodology. Some authors spent considerable time and energy going well beyond their oral or contractual presentation to provide a quality assessment of their softwares.

Though the numerical optimization field is changing very rapidly we have to communicate the capabilities of the numerical optimization techniques and the areas of application of automatic design, even knowing that some of the methodologies presented in this volume will become outdated soon.

The ability to apply numerical optimization techniques to real life problems has improved significantly over the last ten years, and this workshop would not have been possible without the practical expertise of several scientific and industrial partners who have adapted numerical optimization techniques to their own daily problems. Some of those industrial applications are available in the final report of the AERO-89-0026 contract.

We would like to express our particular thanks to D. Knörzer of the European Comission DG-XII for his constant technical interest and fruitful discussions during the two years contract period.

We wish to express our gratitude to the faculty and staff of Universitat Politècnica de Catalunya who contributed to the success of the workshop and gave access to the contributors to the FLAVIA graphic software and made real time presentations on the workstations.

We also thank AERO-89-0026 partners for providing contributions which complied "almost perfectly" to the rules imposed by the coordinator and, in particular, F. Beux, A. Dervieux, D. Joannas, B. Mantel, J. Miller, K. Papailiou, P. Perrier, O. Pironneau, M. Ravachol, H. Schwarten and V. Selmin for interesting discussions and helpful advices during the workshop and during the two years contract period.

The editors thank in particular partners from Daimler-Benz Aerospace Airbus, ALENIA and Dassault Aviation industries, and INRIA, NLR and NTUA institutions for their help and involvement in the definition of the workshop test cases and output formats with great enthusiasm and professional care.

The editors are grateful to Prof. Dr. E. H. Hirschel as the general editor of the "Notes on Numerical Fluid Mechanics" and to the staff of Vieweg-Verlag for the opportunity to publish the results of the EUROPT workshop in this series.

Special thanks are also due to S. Gosset and A. Patry for their careful preparation of the document. We hope these proceedings will be used as a classic reference by young scientists and engineers in the years to come.

Jacques Périaux
Gabriel Bugeda
Panagiotis K. Chaviaropoulos
Theo E. Labrujere
Bruno Stoufflet

December 1994

CONTENTS

1. INTRODUCTION

Despite progress toward automated shape design in industry has been penalized until now by excessive computing costs, useful innovative design methodologies have been proposed (see [1],[2],[3],[4]). These new methodologies can be used for the computation of different academic and industrial designs like nozzles, airfoils and wing-body configurations with inviscid flows modeled by the (full) potential and Euler equations.

Since the designer has a precise idea of the pressure distribution that will produce the desired performance, not only optimization problems, but also inverse problems have to be considered in current design. The main differences between optimization and inverse problems are the following:

- The final goal of optimization problems is to minimize an objective function which is a measure of the quality of the design. Better designs are considered to produce a smaller value of the objective function than worst designs. Typical objective functions are expressed in terms of drag and, in this case, the objective is to minimize drag.
- The final goal of inverse problems is to get a design producing a pressure distribution as close as possible to a given one under specific flow conditions. In this case the pressure distribution is known from the desired aeronautic performance and the objective is to get the shape of the design producing, as approximately as possible, that distribution.

  In order to check the quality of the algorithms for the resolution of inverse problems some reconstruction problems have been proposed for the workshop. In this case, the given pressure distribution corresponds to a well known geometry. Then the algorithms must be able to provide exactly that geometry as the solution of the problem.

In both types of problems, specially in optimization ones, some restrictions can be added to the problem. This restrictions fix minimum or maximum values of some specific characteristics of the design. A typical example is the lift that, normally, must be maintained over a minimum value to assure an adequate aeronautic behavior.

The goal of this volume is to describe methodologies and algorithms - solvers, optimizers and their integration - implemented by seven partners of the European AERO 89-0026 project. The validation process of the design softwares in terms of accuracy and efficiency is presented in the workshop. The solution results of contributors obtained on a selection of nine test cases dealing with optimization and inverse problems are analyzed and discussed.

Chapter 2. "Definition of the problems for analysis" describes a little bit how the workshop test cases were defined, and contains a detailed description of all the workshop test cases.

Chapter 3. "Contributions to the resolution of the workshop test cases" contains the different contributions to the workshop from the partners of the AERO 89-0026 project. This chapter is structured in a partner by partner basis. For each partner there is, first, a theoretical description of the used methodology and, second, the description of the results of the resolution of some of the proposed test cases.

Chapter 4. "Description of the graphic software used for the workshop" describes the graphic facilities that were available for the presentation and comparison of the different contributions during the workshop.

Chapter 5. "Synthesis of the workshop test cases" makes a revision of all the contributions presented to the workshop. For each test case the following points are revised:

- Main characteristics and the reasons that make that test case interesting for the workshop.
- Difficulties for the resolution of the test case.
- Possible interest of the test case for future workshops to be held in the context of future projects like ECARP (European Computational Aerodynamics Research Project).
- Synthesis and comparison of the results provided by different contributors.

Chapter 6. "Conclusions and further comments" closes this volume with the main conclusions and comments that can be extracted from the results presented during the workshop.

REFERENCES

[1] Jameson, A. *Computational Algorithms for Aerodynamic Analysis and Design*, Contribution to the INRIA 25th Anniversary Conference on Computer Science and Control, December 1992.

[2] Beux, F. and Dervieux A. *Exact gradient shape optimization of a 2-D Euler flow* Finite Elements in Analysis and Design 12 (1992), p.p. 281-302.

[3] Huffman, W. P., Melvin, R. G., Young D. P., Johnson, F. T., Bussoletti, J. E., Bieterman, C. L. and Hilmes, C. L. *Practical Design and Optimization in Computational Fluid Dynamics* 24th Fluid Dynamics Conference, Orlando, FI, AIAA 93-3111

[4] Salas, M., Ta'asan, S. and Kuruvila, G. *Aerodynamic Design and Optimization in One-Shot* 28th Aerospace Sciences Meeting and Exhibit, paper AIAA 92-0025

2. DEFINITION OF THE PROBLEMS FOR THE ANALYSIS

During the last years with a constant increasing activity, an increasing number of scientific meetings gathering expert researchers and engineers have taken place in the field of Computational Fluid Dynamics.

Among these events, a class of conferences called workshops are of particular interest in providing opportunities to validate numerical methodologies by comparisons of results obtained by different methods or by direct comparisons of computational and experimental results on flow problems defined as accurately as possible. The outcome of those workshops are of the first importance in the improvement of numerical simulation codes.

The workshop organized within the Brite-Euram project "Optimum Design in Aerodynamics" (AERO-0026C contract) started in June 90, fits in with this class of meetings with an even more ambitious perspective, named the validation of Optimum Design methods, coupling flow analysis solvers and optimizers.

In that direction, the partners of the project have agreeded to define a number of test cases to promote comparisons in terms of accuracy and efficiency of shape design computations with their methodologies. These test cases include three types of problems:

i) *Reconstruction* test cases where the pressure distribution (or another quantity distribution) for a given configuration, once calculated for a given operational condition, serves as design criterion to reconstruct this configuration from an initial guess. An important point of this type of test case is that the exact solution is known.

ii) *Inverse* test cases where a given pressure distribution (or another quantity distribution) serves as design criterion to calculate an optimal configuration corresponding to this pressure distribution. In this case the optimum design will be that providing the pressure distribution closest to the target one.

iii) *Optimization* problems with possible constraints (linear or non linear) to design shapes that correspond to criteria minimization.

The flows under consideration in this workshop are governed by incompressible or compressible full potential equations (inviscid irrotational flows) or Euler equations (inviscid). Two test cases involve the coupling of full potential and boundary layer equations to simulate viscous flows.

The partners defined 17 different test cases from which 9 were finally selected. Table 1 gives a synthesis of the final 9 test cases together with the corresponding participating partners. The main characteristics of the selected test cases are the following:

Test case 1 (T1): 2-D half nozzle reconstruction problem (subsonic).

Test case 2 (T2): 2-D half nozzle optimization problem (subsonic).

Test case 3 (T3): 2-D half nozzle optimization problem (transonic).

Test case 4 (T4): Korn airfoil reconstruction problem (shockless transonic).

Test case 6 (T6): Single point RAE2822 airfoil drag minimization problem (transonic).

Test case 8 (T8): Two point NACA4412 airfoil inverse design with viscous corrections (subsonic and transonic targets).

Test case 10 (T10): 2-D Williams two element configuration reconstruction problem (subsonic).

Test case 14 (T14): Single point DLR-F4 wing (subsonic).

Test case 17 (T17): 3-D double turning nozzle reconstruction problem (subsonic).

Table 1: Selected test cases

Nr.	Problem type	Flow case	Case Defined by	Interested Partners
T1	Inverse Problem M=0.2	Nozzle subsonic Inviscid	INRIA/NTUA	NTUA INRIA UPC TRITECH
T2	Optimization Prob. M=0.2	Nozzle subsonic Inviscid	INRIA/NTUA	NTUA INRIA UPC
T3	Optimization Prob. M=0.5	Nozzle transonic Inviscid	INRIA/NTUA	NTUA INRIA UPC
T4	Reconstruction Prob. M=0.75 $\alpha = 0^{\underline{o}}$ NACA64A410 $\rightarrow$ Korn	Single Airfoil Subsonic Inviscid + Viscous	DASSAULT	NLR UPC ALN INRIA NTUA DASSAULT DA
T6	Optimization Prob. M=0.73 $\alpha = 2^{\underline{o}}$ RAE2822 $\rightarrow$? for given C_L (=1.05)	Single Airfoil Subsonic Inviscid	DASSAULT	ALN INRIA DASSAULT
T8	Inverse Prob. Multi-point M=0.2 and M=0.77 NACA4412 $\rightarrow$?	Single Airfoil Transonic Viscous	NLR	ALN NLR DASSAULT
T10	Inverse Prob.	Two-element airfoil Incompressible Inviscid	DA	DA UPC DASSAULT
T14	Inverse Prob. M=0.3	3D Wing Inviscid	NLR	DA NLR NTUA
T17	Reconstruction Prob. M=0.2	3D Nozzle Subsonic Inviscid	NTUA	INRIA NTUA DASSAULT

The above test problems were chosen because of their particular features:

i) They should be simple 2-D/3-D analytical or well referenced geometries well suited for comparisons of a wide range of solution and optimization methods (integral methods, finite differences, finite volumes or finite elements).

ii) Integrated accurate flow analysis solvers and efficient optimizers should be used to compute feasible shapes efficiently from the original ones.

iii) Test cases should not include at this stage laminar Navier Stokes flow solvers which still remain costly for shape optimization problems on conventional computers.

iv) Existence of well documented results for T4, T10 and T14 problems.

The main challenge of the workshop was the assessment of the optimization methods in the search of shapes (reconstruction or optimization design) with subsonic and/or transonic inviscid flows including possible viscous corrections.

2.1 TEST CASE T1:
INVERSE PROBLEM FOR A NOZZLE

Design conditions

The shape of the target nozzle is represented in Figure 1. The total length of the zone on interest is 2.0, i.e., x varies between 0 and 2, it is a convergent-divergent nozzle with symmetry with respect to $x = 1$. Height is 0.5 at $x = 0.0$ and $x = 2.0$, and 0.25 at $x = 1.0$ The bottom shape is a straight segment. The top shape is the following sine curve:

$$y_t(x) = 0.373 + 0.125 sin(\pi(x - 1.5)) .$$

Flow regime: subsonic with Mach number equal to 0.2 at farfield. The computational domain is extended for positive, $x = 4$, and negative, $x = -2$, in order that the solution be independent of the domain extension.

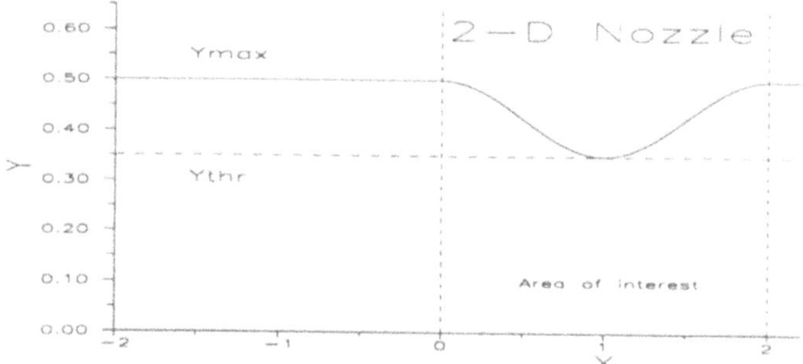

Figure 1. Analysis domain for the T1 test case

Initial conditions

The initial shape is a straight duct defined as follows:

$$y_t(x) = 0.5 .$$

Given data for inverse solution

The following function is given:

$$p(x) = pressure(x, y_t) \quad 0 \le x \le 2 .$$

This data will be obtained from the results of a direct calculation performed for the target sine shape.

2.2 TEST CASE T2: SUBSONIC OPTIMIZATION PROBLEM FOR A NOZZLE

The data conditions of this problem are identical to the ones for test case T1 except the following points:

Constraints

$$y_t(0) = y_t(2) = 0.5$$
$$y_t(1) = 0.25$$

and y_t is a univoque function of x for $0 \leq x \leq 2$.

Flow regime

Mach number equal to 0.2 (subsonic) at farfield. The computational domain should be extended as in test case T1.

Optimization problem

Find a shape y_t which minimize the following cost function:

$$\int_0^2 \left| \frac{\partial p}{\partial x}(x, y_t(x)) \right|^\alpha dx$$

where p is (again) the pressure, and $\alpha = 2$ or $\alpha = 4$ under the constraint $y_t(1) = 0.25$.

2.3 TEST CASE T3: TRANSONIC OPTIMIZATION PROBLEM FOR A NOZZLE

This test case is identical to T2, but the Mach number at farfield is 0.5 (transonic).

2.4 TEST CASE T4: INVERSE PROBLEM FOR A 2D SINGLE AIRFOIL

Design conditions

It is proposed to recover the Korn airfoil for subsonic inviscid flow at Mach number

0.75 and at angle of attack 0º. Table 2 contains the pointwise geometrical definition of the Korn airfoil.

The target pressure is obtained by direct calculation around the Korn airfoil with susmentioned freestream conditions.

Initial conditions

It is proposed to start from the NACA64A410 with the above flow conditions. The definition of the NACA64A410 profile is provided in Table 3.

2.5 TEST CASE T6: OPTIMIZATION PROBLEM FOR A 2D SINGLE AIRFOIL

One point design

A drag minimization problem issued from reference [1] is proposed. The chosen initial profile is the RAE2822 airfoil. The freestream Mach number is taken to be $M = 0.730$ and the initial angle of attack is $\alpha = 2^{\circ}$. The flow is modeled by the full potential or the Euler equations. Two minimization problems are proposed.

Improvement of existing profile via an inverse problem

Firstly, we formulate the problem as a perturbation of an inverse problem. The target pressure distribution p_{tar} is taken to be the actual pressure distribution of the initial profile predicted by the numerical solution of the flow equations. The inclusion of this target distribution will force the method to generate a profile with a lift coefficient close to that of the initial profile. Drag coefficient C_D is added to the cost function so that the expression of the cost function I is:

$$I = I_{tar} + \beta C_D$$

where I_{tar} is the cost function corresponding to the pure inverse problem. The addition of a drag penalty now causes the method to reshape the profile to reduce its drag where β is a parameter. It is asked to evaluate the sensitivity of the minimization procedure and of the solution to the magnitude of this parameter.

Improvement of existing profile via a minimization

Secondly, we consider the previous problem as a pure minimization problem expressed as:

Minimize $I = C_D$ under the constraint of a given lift C_L.

Table 2: Definition of Korn airfoil

Korn airfoil			
Upper side		Lower side	
X	Y	X	Y
-0.000062	0.003256	-0.000062	0.003256
0.000515	0.007246	0.000152	-0.000731
0.002054	0.010923	0.001306	-0.004583
0.004862	0.014394	0.003513	-0.008207
0.008925	0.017870	0.006801	-0.011672
0.014103	0.021385	0.011244	-0.014932
0.020337	0.024907	0.016879	-0.018082
0.027583	0.028402	0.023659	-0.021163
0.035808	0.031830	0.031534	-0.024183
0.044989	0.035155	0.040460	-0.027131
0.055095	0.038336	0.050400	-0.029992
0.066103	0.041318	0.061321	-0.032746
0.078015	0.044007	0.073191	-0.035373
0.090953	0.046354	0.085982	-0.037856
0.105001	0.048469	0.099663	-0.040176
0.120082	0.050462	0.114209	-0.042317
0.136112	0.052348	0.129590	-0.044265
0.153029	0.054129	0.145777	-0.046009
0.170776	0.055802	0.162740	-0.047540
0.189300	0.057360	0.180446	-0.048852
0.208550	0.058801	0.198858	-0.049939
0.228472	0.060118	0.217938	-0.050800
0.249016	0.061308	0.237643	-0.051428
0.270128	0.062366	0.257932	-0.051821
0.291754	0.063289	0.278759	-0.051974
0.313840	0.064072	0.300077	-0.051882
0.336330	0.064713	0.321841	-0.051539
0.359167	0.065208	0.344007	-0.050938
0.382294	0.065555	0.366531	-0.050074
0.405653	0.065749	0.389370	-0.048942
0.429184	0.065788	0.412485	-0.047539
0.452828	0.065667	0.435835	-0.045864
0.476526	0.065382	0.459385	-0.043916
0.500217	0.064928	0.483100	-0.041700
0.523841	0.064297	0.506947	-0.039222
0.547340	0.063479	0.530896	-0.036495
0.570659	0.062463	0.554916	-0.033533
0.593747	0.061232	0.578984	-0.030356
0.616562	0.059772	0.603077	-0.026994
0.639072	0.058072	0.627172	-0.023484
0.661248	0.056130	0.651250	-0.019878
continue on the next page			

Table 2 bis: Definition of Korn airfoil (continuation)

Korn airfoil (cont.)			
Upper side		Lower side	
X	Y	X	Y
0.683064	0.053952	0.675284	-0.016242
0.704491	0.051552	0.699239	-0.012662
0.725501	0.048945	0.723058	-0.009238
0.746064	0.046153	0.746653	-0.006078
0.766148	0.043201	0.769900	-0.003284
0.785720	0.040115	0.792640	-0.000928
0.804748	0.036925	0.814704	0.000958
0.823195	0.033667	0.835929	0.002386
0.841024	0.030373	0.856176	0.003392
0.858199	0.027085	0.875331	0.004028
0.874676	0.023843	0.893306	0.004349
0.890408	0.020689	0.910028	0.004407
0.905344	0.017668	0.925436	0.004252
0.919425	0.014820	0.939482	0.003929
0.932589	0.012179	0.952123	0.003482
0.944768	0.009777	0.963326	0.002952
0.955897	0.007633	0.973062	0.002379
0.965909	0.005760	0.981311	0.001800
0.974742	0.004164	0.988059	0.001253
0.982338	0.002846	0.993299	0.000773
0.988643	0.001799	0.997031	0.000393
0.993611	0.001010		
0.997201	0.000467		
1.000000	0.000056		

Multi point design

The optimization procedure is not limited to a single design point. The above minimization problem can be extended to multipoint criteria design. The RAE2822 is the proposed initial profile with three design targets ($C_L = 1.05$ at $M_\infty = 0.73$), ($C_L = 0.95$ at $M_\infty = 0.74$), ($C_L = 0.85$ at $M_\infty = 0.75$). For this particular case, the criteria is the sum of cost functions for several design points.

References

[1] Jameson, A. *Automatic Design of Transonic Airfoil to Reduce the Shock-Induced Pressure Drag*, 31st Israel Annual Conference Aviation and Aeronautics, February 21-22, 1990.

Table 3: Definition of NACA64A410 airfoil

NACA64A410 airfoil			
Upper side		Lower side	
X	Y	X	Y
0.000000	0.000000	0.000000	0.000000
0.003500	0.009020	0.006500	-0.006780
0.005820	0.011120	0.009180	-0.007960
0.010590	0.014510	0.014410	-0.009690
0.022760	0.020950	0.027240	-0.012510
0.047490	0.030340	0.052510	-0.015920
0.072300	0.038650	0.077700	-0.019190
0.097370	0.043800	0.102630	-0.019960
0.147480	0.053660	0.152520	-0.022440
0.197700	0.061260	0.202300	-0.024060
0.248000	0.067050	0.252000	-0.024990
0.298340	0.071310	0.301660	-0.025370
0.348710	0.074140	0.351290	-0.025180
0.399100	0.075520	0.400900	-0.024360
0.449500	0.075220	0.450500	-0.022660
0.499890	0.073440	0.500110	-0.020240
0.550250	0.070400	0.549750	-0.017360
0.600570	0.066240	0.599430	-0.014180
0.650850	0.061060	0.649150	-0.010860
0.701080	0.054900	0.698920	-0.007600
0.751260	0.047800	0.748740	-0.004600
0.801510	0.039670	0.798490	-0.002290
0.851480	0.030180	0.848520	-0.001320
0.901040	0.020380	0.898960	-0.000760
0.950530	0.010280	0.949470	-0.000480
1.000000	0.000210	1.000000	-0.000210

2.6 TEST CASE T8: TEST CASE FOR MULTI-POINT 2D AIRFOIL DESIGN

For multi-point airfoil design a test case is defined aiming at the design of an airfoil which combines a favourable high speed and a favourable low speed performance. Analogous problems will be encountered at e.g. the design of a transport aircraft outer wing section (no flap, no slat) or at the design of a helicopter rotor blade section.

The formulation of the design criteria is essential for a realistic design test case. Two target pressure distributions have been chosen. Target pressure distribution 1 has been chosen to produce a favourable high lift capacity. Target pressure 2 has been chosen for its favourable high speed performance (weak shock). A complete definition of the test case is given by means of Table 4. Here, the initial geometry is specified in a number of contour points. x is the coordinate measured along the chord of the airfoil, where the chord is defined as the line connecting the trailing edge point with the point at the nose which has the largest distance to the trailing edge point. The target pressure distributions are specified pointwise as a function of x.

Table 4: Definition of T8 test case

Initial geometry		Target 1	Target 2
NACA 4412		$\alpha = 7.8^{\circ}$ $M = 0.2$ $Re = 5 * 10^6$	$\alpha = 0.0^{\circ}$ $M = 0.77$ $Re = 10 * 10^6$
X	**Y**	C_p	C_p
1.0	-0.0013	0.20	0.210
0.95	-0.0016	0.24	0.265
0.9	-0.0022	0.26	0.300
0.8	-0.0039	0.26	0.293
0.7	-0.0065	0.23	0.193
0.6	-0.0100	0.18	0.050
0.5	-0.0140	0.14	-0.128
0.4	-0.0180	0.15	-0.395
0.3	-0.0226	0.23	-0.460
0.25	-0.0250	0.29	-0.385
0.2	-0.0274	0.38	-0.295
0.15	-0.0288	0.48	-0.195
0.1	-0.0286	0.61	-0.070
0.075	-0.0274	0.69	0.000
0.05	-0.0249	0.78	0.100
0.025	-0.0195	0.92	0.230
0.0125	-0.0143	0.99	0.400
0.0	0.0	0.00	1.010
0.0125	0.0244	-3.26	-0.500
0.025	0.0339	-3.26	-0.970
0.05	0.0473	-3.24	-1.040
0.075	0.0576	-3.21	-1.022
0.1	0.0659	-3.14	-1.002
0.15	0.0789	-3.00	-0.971
0.2	0.0880	-2.77	-0.941
0.25	0.0941	-1.90	-0.914
0.3	0.0976	-1.32	-0.889
0.4	0.0980	-0.80	-0.842
0.5	0.0919	-0.51	-0.799
0.6	0.0814	-0.30	-0.760
0.7	0.0669	-0.14	-0.435
0.8	0.0489	-0.01	-0.290
0.9	0.0271	0.10	-0.065
0.95	0.0147	0.14	0.073
1.0	0.0013	0.20	0.210
Constraints		chord=1 trailing edge angle $\geq 10^{\circ}$ thickness t/c=11%	
Weight factors		0.5	0.5

The angle of attack α with respect to the chord of the airfoil may be kept fixed or may be considered as a design parameter. The actual two-point design problem comes down to the determination of the shape of an airfoil such that

$$\sum_{i=1}^{2} \int_{x=0}^{x=1} [C_P^i - C_{P_{tar}}^i]^2 dx$$

is minimized. Target 1 should be produced by the designed airfoil at M=0.2 ($\alpha = 7.8^{\underline{o}}$), target 2 should be produced at M=0.77 ($\alpha = 0^{\underline{o}}$).

By chosing only one of the pressure distributions as a target, the test case is reduced to a single-point design problem.

The geometric constraints defined in Table 4 may be completed with other constraints in order to prevent unrealistic geometries such as negative thickness at the trailing edge.

2.7 TEST CASE T10: RECONSTRUCTION PROBLEM FOR A 2D MULTI-ELEMENT AIRFOIL

Design conditions

This test case deals with the reconstruction of a two-element configuration from their pressure distribution. Table 5 contains the pointwise geometrical definition of the target profiles and its pressure distribution which correspond to the, so called, Williams two-elements airfoil. The pressure distribution has been taken from [1], and it has been obtained by using a special conformal mapping (see [1]) with $M = 0.0$ and the onset flow in the x-direction. The target pressure distribution for this test case can be the one contained in Table 5, or a different one obtained from a direct analysis of the Williams airfoil.

Known geometric properties of the profiles to be designed

Each of the two elements is free to rotate around its leading edge. These are fixed relativ to the global $(x - y)$ frame. The position of the leading edges x^{LE}, y^{LE} relativ to the global $(x - y)$ frame and the chord lengths L_1 and L_2 are specified. We are free to set the value of L_1 to 1, which means that all lengths are normalized with the chord length of the first element.

The starting design is defined as two NACA0012 airfoils placed horizontally.

leading edge (first element): $x_1^{LE}/L_1 = 0.0 \quad y_1^{LE}/L_1 = 0.0$

leading edge (second element): $x_2^{LE}/L_1 = 0.970804 \quad y_2^{LE}/L_1 = 0.070366$

ratio of chord length: $L_2/L_1 = 0.367807$.

Table 5: Definition of the two-elements Williams airfoil

Williams airfoil					
first element			second element		
X	Y	C_P	X	Y	C_P
1.000000	-0.026320	1.000000	1.333520	-0.134080	1.000000
0.999090	-0.026020	0.581980	1.333160	-0.133940	0.486060
0.992380	-0.023950	0.677240	1.330070	-0.132610	0.468620
0.979920	-0.020840	0.754440	1.323930	-0.130130	0.480110
0.962390	-0.017960	0.787030	1.314690	-0.126710	0.495800
0.940070	-0.016460	0.776500	1.302360	-0.122590	0.509960
0.912950	-0.016920	0.734630	1.287080	-0.118080	0.520430
0.881030	-0.019380	0.676310	1.269070	-0.113470	0.526230
0.844540	-0.023530	0.613260	1.248680	-0.109040	0.526860
0.803950	-0.028920	0.552020	1.226360	-0.105040	0.522090
0.759840	-0.035090	0.495250	1.202640	-0.101610	0.511820
0.712870	-0.041670	0.443720	1.178080	-0.098830	0.496040
0.663720	-0.048280	0.397550	1.153260	-0.096670	0.474830
0.613040	-0.054630	0.356780	1.128720	-0.095040	0.448370
0.561460	-0.060450	0.321610	1.104950	-0.093790	0.416960
0.509550	-0.065510	0.292450	1.082400	-0.092730	0.381080
0.457900	-0.069360	0.269960	1.061410	-0.091690	0.341410
0.407010	-0.072660	0.255000	1.042270	-0.090490	0.298860
0.357390	-0.074470	0.248690	1.025200	-0.089020	0.254600
0.309500	-0.074990	0.252360	1.010370	-0.087180	0.210110
0.263770	-0.074180	0.267670	0.997890	-0.084950	0.167320
0.220620	-0.072040	0.296580	0.987820	-0.082350	0.129440
0.180420	-0.068590	0.341480	0.980170	-0.079470	0.104700
0.143530	-0.063900	0.405220	0.974900	-0.076420	0.123880
0.110260	-0.058060	0.491050	0.971940	-0.073350	0.309800
0.080920	-0.051200	0.602010	0.971140	-0.070390	0.804030
0.055760	-0.043460	0.738560	0.972280	-0.067660	0.996090
0.035030	-0.035010	0.889390	0.975060	-0.065250	0.880610
0.018940	-0.026010	0.997220	0.979170	-0.063230	0.768710
0.007670	-0.016640	0.838000	0.984280	-0.061590	0.694820
0.001380	-0.007060	-0.224890	0.990130	-0.060270	0.643460
0.000180	0.002600	-2.288980	0.996530	-0.059140	0.598920
0.004090	0.012260	-3.283880	1.003370	-0.058080	0.548130
0.013030	0.021790	-3.136480	1.010640	-0.056920	0.479880
0.026870	0.031060	-2.771640	1.018400	-0.055600	0.385390
0.045430	0.039890	-2.454460	1.026750	-0.054070	0.259700
0.068520	0.048110	-2.207720	1.035800	-0.052370	0.103050
0.095890	0.055540	-2.014890	1.045650	-0.050580	-0.079000
0.111110	0.058920	-1.933320	1.056370	-0.048850	-0.276690
0.144450	0.064890	-1.791720	1.067980	-0.047300	-0.478040
0.181390	0.069720	-1.671180	1.080460	-0.046110	-0.670910
continue on the next page					

Table 5 bis: Definition of the two-elements Williams airfoil (continuation)

Williams airfoil (cont.)					
first element			second element		
X	Y	C_P	X 1	Y	C_P
0.221620	0.073320	-1.565250	1.093770	-0.045400	-0.844610
0.264780	0.075630	-1.469760	1.107810	-0.045310	-0.990800
0.310510	0.076600	-1.382020	1.122490	-0.045940	-1.103880
0.358390	0.076230	-1.300380	1.137660	-0.047370	-1.180890
0.407980	0.074570	-1.223920	1.153210	-0.049630	-1.221150
0.458840	0.071680	-1.152260	1.168970	-0.052760	-1.225860
0.510480	0.067650	-1.085420	1.184810	-0.056730	-1.197650
0.562370	0.062630	-1.023760	1.200560	-0.061520	-1.140160
0.613960	0.056760	-0.967910	1.216100	-0.067040	-1.057640
0.664690	0.050230	-0.918760	1.231260	-0.073220	-0.954690
0.713930	0.043210	-0.877380	1.245920	-0.079920	-0.836000
0.761060	0.035870	-0.844940	1.259940	-0.087000	-0.706100
0.805410	0.028370	-0.822240	1.273190	-0.094290	-0.569250
0.846320	0.020810	-0.808610	1.285530	-0.101610	-0.429270
0.883150	0.013250	-0.799100	1.296830	-0.108730	-0.289450
0.915350	0.005670	-0.778530	1.306960	-0.115440	-0.152420
0.942520	-0.001970	-0.713610	1.315750	-0.121490	-0.020020
0.964500	-0.009560	-0.556770	1.323030	-0.126630	0.107020
0.981310	-0.016690	-0.286150	1.328610	-0.130610	0.230020
0.992970	-0.022490	0.052240	1.332230	-0.133180	0.356840
0.999160	-0.025860	0.393900	1.333520	-0.134080	1.000000
1.000000	-0.026320	1.000000			

References

[1] Williams, B. R. "An exact test case for the plane potential flow about two adjacent lifting airfoils" ARCR& M, 3717, 1973.

2.8 TEST CASE T14: DESIGN AND RECONSTRUCTION TEST CASE FOR SINGLE-POINT 3D WING DESIGN

This test case is related to transport aircraft wing design.The objective of this design exercise is to improve the (inviscid) low speed high lift pressure distribution for a known wing-body configuration by modifying wing section contours, while retaining the body shape, the planform of the wing and the wing root section.

Design case

The starting point is the well known F4 configuration (see Figures 1,2) of which the body geometry and the wing geometry will be specified by means of a number of contour points. The target pressure distribution will be specified by means of chordwise

Figure 1: The F4 configuration

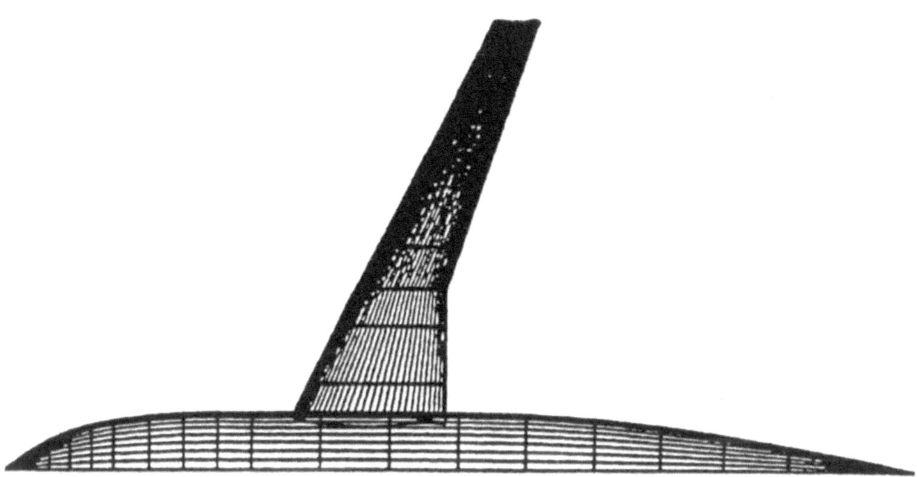

Figure 2: F4 planform

pressure distributions in a number of wing sections.

The target pressure distribution has been obtained by modifying the pressure distribution on the wing calculated by means of the NLR potential flow solver (see references [1],[2]) for $C_L \approx 1.60$. This pressure distribution has been modifyed to reduce the high supersonic velocity peaks to subsonic values, while retaining approximately the sectional lift coefficients. Figure 3 presents the calculated upper surface pressure distributions together with the modified (target) pressure distributions. The lower surface pressure distribution should remain unaltered.

The application of geometric constraints might appear too restrictive in this de-

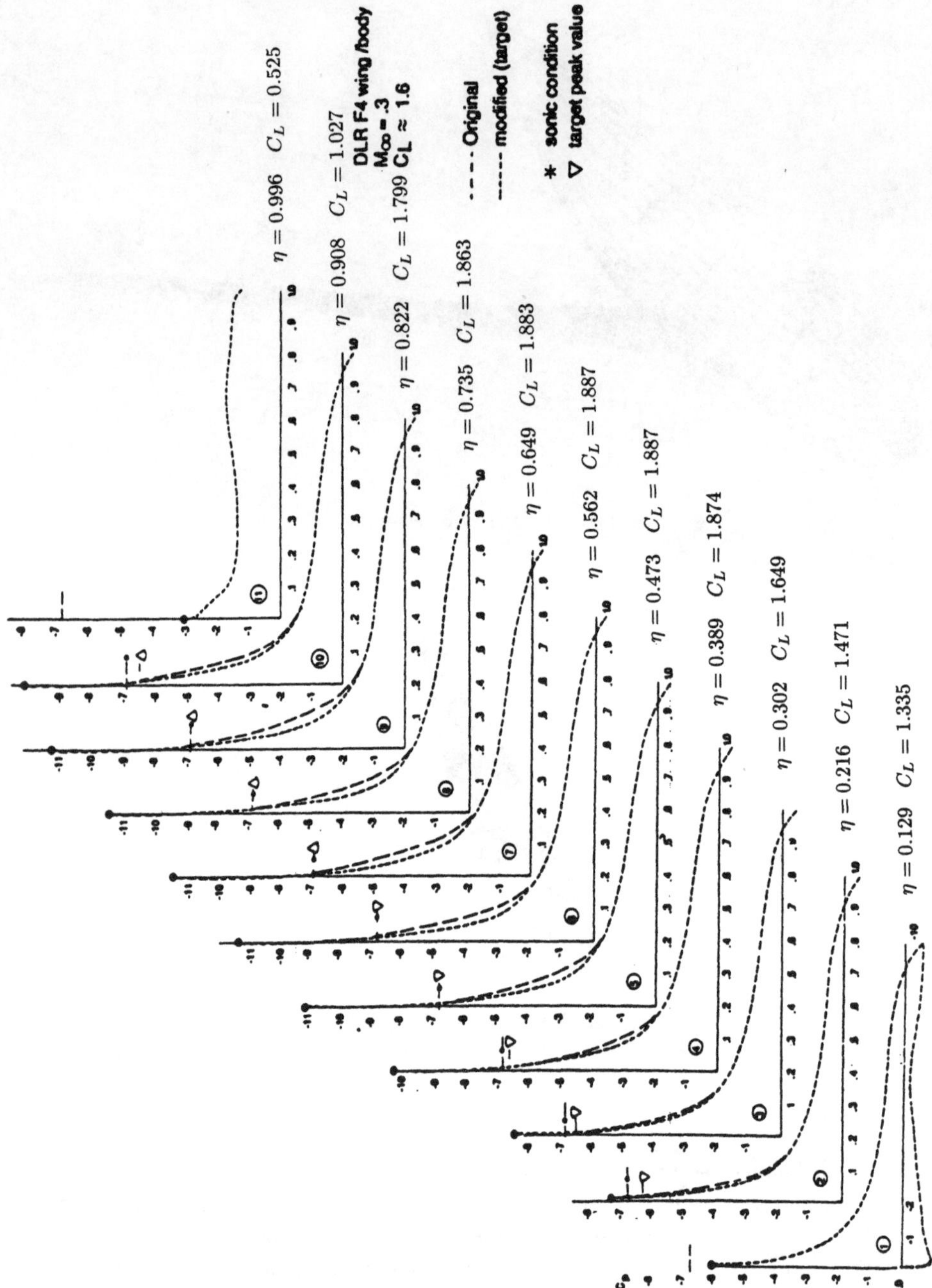

Figure 3: Target (upper surface) pressure distributions for 3D wing design

sign exercise, therefore it is suggested to consider this as an option. The constraint that may be applied in this case is the retainment of the wing thickness at 20% of the chord (front spar).

Reconstruction case

The F4 wing has also been chosen for a 3D reconstruction test case. The initial configuration includes the F4 body, the F4 planform and the F4 wing root section, but it has a perturbed wing geometry in comparison with the F4 (perturbation in terms of camber and thickness).

Starting with the initial configuration, the geometry of the F4 configuration should be reconstructed according with the following steps:

- Calculation of the pressure distribution on the F4 wing-body configuration for $\alpha = 6^{\underline{o}}$ and $M = 0.3$
- Application of the design algorithm with the pressure distribution from the first step as target and with the initial configuration as starting point.

References

[1] Voorem, J. van der, Wees, A. J. van der, and Meelker, J. H. "Transonic potential flow calculations about transport aircraft" AGARD Conf. Proc. No. 412 (1986).

[2] Voorem, J. van der, Wees, A. J., "Inviscid drag prediction for transonic transport wings using a full-potential method" AIAA-90-0576 (1990). Published in tha AIAA Journal.

2.9 TEST CASE T17: RECONSTRUCTION PROBLEM FOR A 3D NOZZLE

Generalities

The present test case is dealing with the reconstruction of a 3D double turning nozzle by the use of the target pressure technique. The flow model which will be used is the irrotational one (incompressible or compressible).

Before the definition of the geometry the following criteria were considered:

- The shape of the nozzle should be fully three-dimensional.
- To avoid data transfer the geometry definition should be analytical and parametrical, so that, each partner can construct the test case by their own means.

Definition of the geometry

The definition of the geometry involves two steps:

a) The definition of the nozzle mean line.

b) The definition of the sections.

The mean line of the nozzle is shown in Figure 4, along with the corresponding parameters:

L_1, L_2, L_3 : lengths in the x, y, z directions

R_1, R_2 : raius on the $(x - z)$ and $(y - z)$ planes respectively

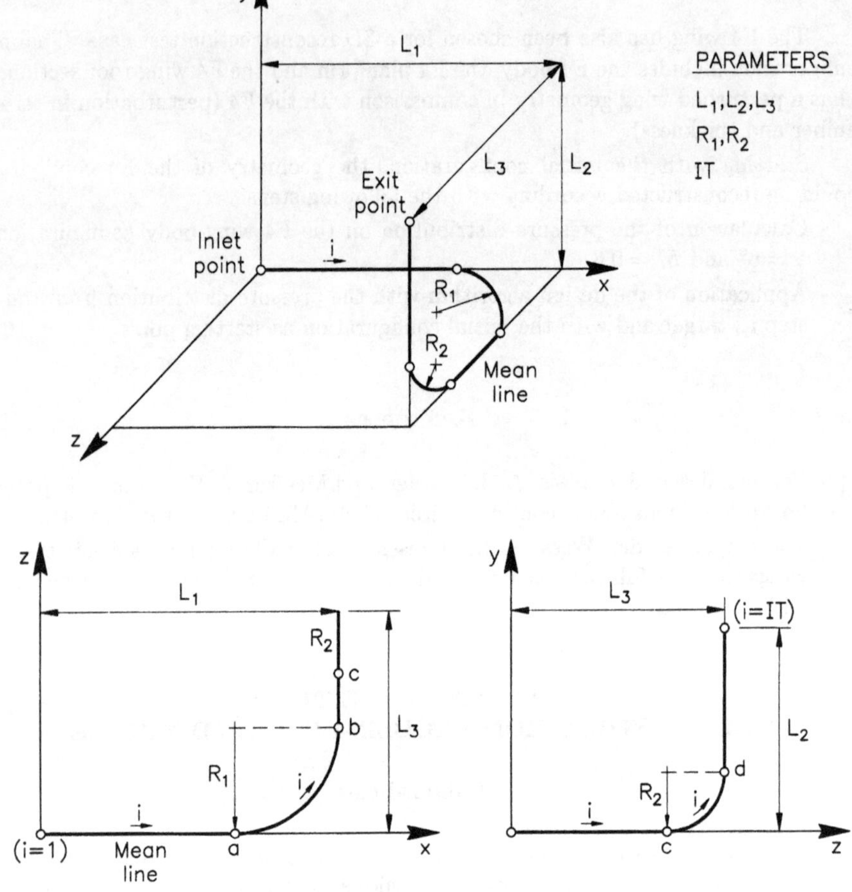

Figure 4: Definition of the mean line

IT : number of equidistant points on the mean line

The cross sections definitions and their evolution are shown in Figure 5, along with the corresponding parameters. The cross sections are lying on planes which are normal to the mean line.

A_1, A_2 : inlet and outlet cross section areas

t : constant cross section aspect ratio

JT, KT :number of equidistant points along j and k directions

The cross section variation from the inlet to the exit plane is considered to be linear:

$$A_i = A_1 + \frac{(i-1)}{(IT-1)}(A_1 - A_2) \, .$$

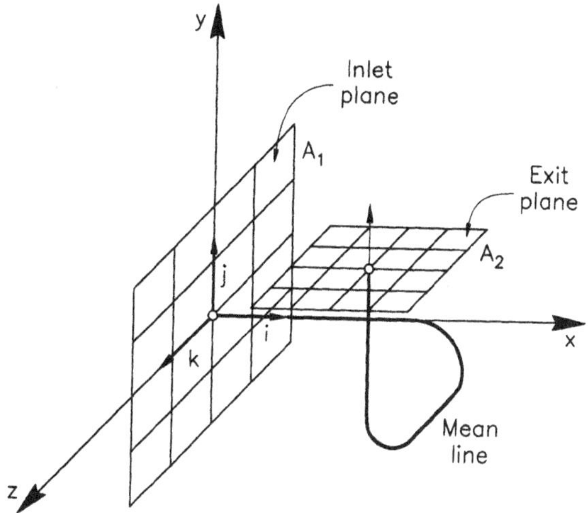

Figure 5: Definition of the cross sections

Application

The grid generation source code is given in pages 21 to 23. Using this code and assigning the following values to the above parameters:

$$(IT, JT, KT) = (30, 7, 7)$$
$$(L_1, L_2, L_3) = (10, 10, 10)$$
$$(A_1, A_2, t) = (25, 9, 1)$$

one may obtain a converging double turning nozzle, the boundary nodes of which are shown in Figure 6.

Test case

The test case proposed here involve the reconstruction of the above geometry for the inlet flow condition $M = 0.2$ (compressible case).

The participants may compute the target pressure boundary distribution by using their direct codes.

Participants using iterative methods are proposed to start with a geometry that maintains the correct "mean line" shape having, however, a constant cross section distribution ($A = A_1$) as initialization. Then the optimization goal will be to recover the cross section law.

Specification for Numerical Results

To check the accuracy of the reconstruction procedure the following residuals are proposed:

$$R_{m1} = \frac{1}{L_1} max\{||r_i - r_i^0||^2\} \quad i = 1, IT$$

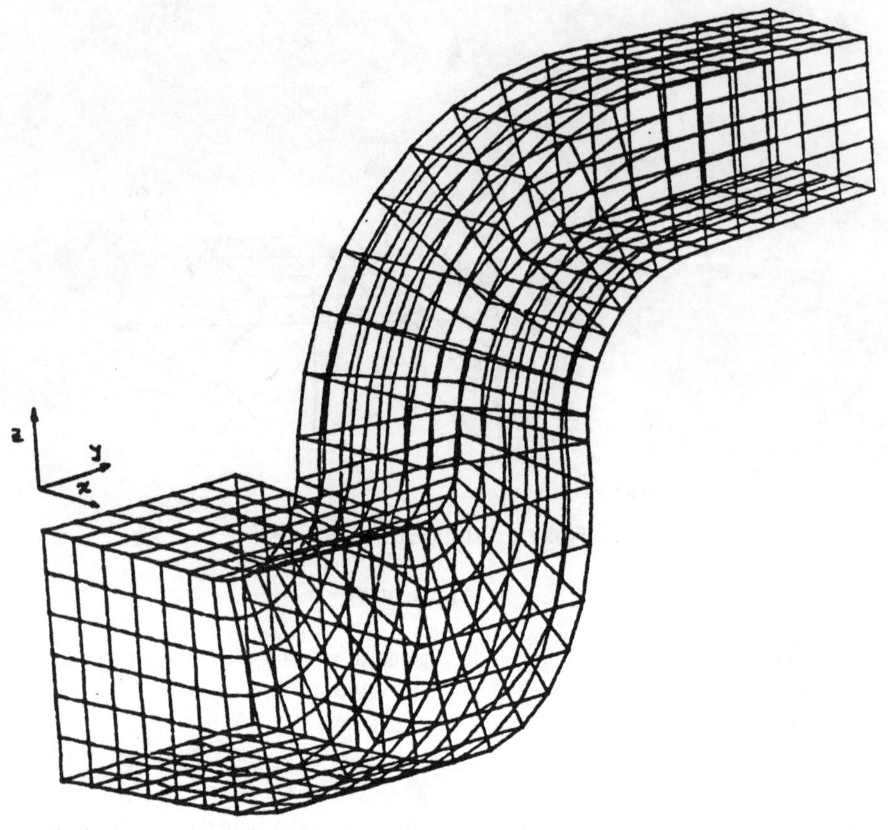

Figure 6: Test case geometry

$$R_{m2} = \frac{1}{IT * L_1}\{\sum_{i=1}^{IT} \|r_i - r_i^0\|^2\}^{\frac{1}{2}}$$

IT : is the mid-line number of points

L_1 : is the characteristic length, here $L_1 = 10$

r_i : is the i-node position vector of the reconstructed grid (topology)

r_i^0 : is the i-node position vector of the original grid (topology)

The mean-line grid nodes are considered to be equidistant.

FORTRAN code

```
c L1,L2,L3 Lengths in x,y,z directions
c R1,R2 Radii
c A1,A2 Inlet and exit areas
c AR Aspect Ratio
c IT,JT,KT Number of grid points in I,J,K, directions

      write(*,*) 'Enter L1,L2,L3'
      read(*,*) l1,l2,l3
      write(*,*) 'Enter R1,R2'
      read(*,*) r1,r2
      write(*,*) 'Enter A1,A2,AR'
      read(*,*) a1,a2,t
      write(*,*) 'Enter IT,JT,KT'
      read(*,*) it,jt,kt

      pi=4.*atan(1.)
      sa=l1-r1
      sb=sa+pi*r1/2.
      sc=sb+l3-(r1+r2)
      sd=sc+pi*r2/2.
      sf=sd+l2-r2

      open(1,file='geowoe',status='unknown')

      do i=1,it

c equidistant points in I direction
      s=sf*(i-1.)/(it-1.)

c linear cross section area variation
      a=a1+(a2-a1)*(i-1.)/(it-1.)
      b=sqrt(a/t)
      h=b*t

c grid steps in J and K directions
      db=b/(jt-1.)
      dh=h/(kt-1.)

c xm.ym,zm mean line coordinates
c xijk,yijk,zijk grid point coordinates

      if(s.le.sa) then
      xm=s
      ym=0.
      zm=0.
      do j=1,jt
        do k=1,kt
          xijk=xm
          yijk=ym-b/2.+(j-1)*db
```

```
    zijk=zm-h/2.+(k-1)*dh
    write(1,100) xijk,yijk,zijk,i,j,k
  enddo
enddo

else if(s.gt.sa.and.s.lt.sb) then
thet=(s-sa)/r1
xm=sa+r1*sin(thet)
ym=0.
zm=r1*(1.-cos(thet))
do j=1,jt
  do k=1,kt
    xijk=xm-(-h/2.+(k-1)*dh)*sin(thet)
    yijk=ym-b/2.+(j-1)*db
    zijk=zm+(-h/2.-(k-1)*dh)*cos(thet)
    write(1,100) xijk,yijk,zijk,i,j,k
  enddo
enddo

else if(s.ge.sb.and.s.le.sc) then
xm=l1
ym=0.
zm=s-sb+r1
do j=1,jt
  do k=1,kt
    xijk=xm-(-h/2.+(k-1)*dh)
    yijk=ym-b/2.+(j-1)*db
    zijk=zm
    write(1,100) xijk,yijk,zijk,i,j,k
  enddo
enddo

else if(s.gt.sc.and.s.lt.sd) then
fi=(s-sc)/r2
xm=l1
ym=r2*(1.-cos(fi))
zm=l3-r2*(1.-sin(fi))
do j=1,jt
  do k=1,kt
    xijk=xm-(-h/2.+(k-1)*dh)
    yijk=ym+(-b/2.+(j-1)*db)*cos(fi)
    zijk=zm-(-b/2.+(j-1)*db)*sin(fi)
    write(1,100) xijk,yijk,zijk,i,j,k
  enddo
enddo

else if(s.ge.sd) then
xm=l1
ym=s-sd+r2
zm=l3
```

```
      do j-1,jt
        do k=1,kt
          xijk=xm-(-h/2.+(k-1)*dh)
          yijk=ym
          zijk=zm-(-b/2.+(j-1)*db)
          write(1,100) xijk,yijk,zijk,i,j,k
        enddo
      enddo

      endif
      enddo

      close(1)
100   format(3(1x,f12.6),3i3)
      stop
      end
```

3. CONTRIBUTIONS TO THE RESOLUTION OF THE WORKSHOP TEST CASES

The following sections contain the different contributions from the consortium of the BRITE/EURAM project "Optimum Design in Aerodynamics" (AERO-89-0026) for the resolution of the workshop test cases previously defined.

Each contribution contains a theoretical description of the optimization techniques used by each partner and the results obtained by using this techniques for the resolution of some of the test cases.

These results were also presented in the workshop on Optimum Shape Design in Aeronautics that took place in Barcelona at the Universitat Politècnica de Catalunya (UPC) on June 1992.

The contributions are ordered according with the following list of partners:

1.- DASSAULT AVIATION
2.- DEUTSCHE AIRBUS
3.- ALENIA
4.- NLR
5.- INRIA ROCQUENCOURT
6.- INRIA SOPHIA ANTIPOLIS
7.- UPC
8.- NTUA

CONTRIBUTION TO PROBLEMS T4 AND T6 FINITE ELEMENT GMRES AND CONJUGATE GRADIENT SOLVERS

Q.V. DINH, B.MANTEL, J. PERIAUX, B.STOUFFLET

Dassault Aviation
78 *quai Marcel Dassault* - 92214 *Saint-Cloud (France)*

SUMMARY

A GMRES solver based on the estimation of the gradient of the objective function gradient algorithm are used for the reconstruction and optimization of airfoils represented by cubic splines and operating at subsonic and transonic conditions. It is shown that the number of iterations has been multiplied by three between the subsonic and the transonic case. For the optimization problem, the drag objective function is penalized to take into account the lift constraint. Two descent methods, namely a fixed step gradient and a conjugate gradient algorithm are investigated. Results are compared with a GMRES optimizer for the solution of the optimality conditions using the calculus of variations with adjoint equations. These methods have been developped in close cooperation with O. Pironneau and A. Vossinis at INRIA Rocquencourt.

INTRODUCTION

The development of a methodology using GMRES as a non linear optimization algorithm used for the design of aerodynamic shapes operating in comressible transonic flow is a useful tool in Engineering. The full potential equation considered in Section 2 is discretized through a Galerkin variational formulation and the numerical computation of boundary integrals performed by adapted Riemann solver. Different parametrizations of the geometry and B-spline representations of the surface and a linear elasticity strategy for unstructured mesh deformation in the field at each iteration of the design procedure are described in Section 3. In Section 4 we first consider reconstruction problems where teh objective is to calculate the profile geometry corresponding to a target pressure distribution and also drag optimization problems subject to aerodynamical constraints. The corresponding GMRES and conjugate gradient algorithms are described for solving the above problems. Numerical results obtained by the above methodologies are discussed in Section 5.

Two different approaches have been considered:

1) development of a methodology using GMRES as an optimization method on the optimality conditions computed via an adjoint equation in close cooperation with INRIA (see INRIA 24 month report for the details of the gradient evaluation).

2) extension of the methodology used for incompressible flow problems avoiding adjoint operator evaluation; this approach is detailed in the sequel.

In a general frame, an optimum shape design problem can be described as follows. The current shape denoted γ belongs to a class of shapes $\mathcal{O}$. The design problem is given by

$$\text{Find a shape } \gamma \text{ such that } \gamma^* = \arg \min_{\gamma \in \mathcal{O}} j(\gamma)$$

where $j(\gamma)$ is the cost function derived from a certain criterion.
In fact, the cost function is given by the following identity

$$j(\gamma) = J(\gamma, \mathbf{W}(\gamma))$$

where $\mathbf{W}(\gamma)$ describes the state of the system i.e that $\mathbf{W}(\gamma)$ is the solution (unique) of the state equation:

$$E(\gamma, \mathbf{W}(\gamma)) = 0.$$

In most of the methods to treat such problems, one usually writes that the optimum γ^* satisfies the optimality condition:

$$\frac{dj(\gamma)}{d\gamma} = 0$$

which is given in our case by

$$\frac{dj(\gamma)}{d\gamma}\delta\gamma = \frac{\partial J}{\partial \gamma}\delta\gamma + < \frac{\partial J}{\partial \mathbf{W}}, \frac{d\mathbf{W}}{d\gamma}\delta\gamma >$$

$$\frac{\partial E}{\partial \gamma} + < \frac{\partial E}{\partial \mathbf{W}}, \frac{d\mathbf{W}}{d\gamma} > = 0.$$

Let define the adjoint state Ψ solution of the adjoint problem

$$\frac{\partial E}{\partial \mathbf{W}}^* \Psi = \frac{\partial J}{\partial \mathbf{W}}$$

then we have

$$\frac{dj(\gamma)}{d\gamma}\delta\gamma = \frac{\partial J}{\partial \gamma}\delta\gamma - < \Psi, \frac{\partial E}{\partial \gamma}\delta\gamma >.$$

This formalism can be viewed equivalent to introducing a Lagrangian. By the introduction of the adjoint problem, the explicit computation of $\frac{d\mathbf{W}}{d\gamma}$ is avoided. The actual computation of the Jacobian of E with respect to the state $\mathbf{W}$ is needed. Its evaluation is directly related to the governing equations of the flow and can lead to extremely cumbersome calculations once the considered model is beyond the one of incompressible inviscid flows [2]. This approach has been investigated in cooperation with INRIA as previously mentionned.

In the study reported here, the derivative is evaluated by the following Gateaux derivation given by

$$\frac{dj(\gamma)}{d\gamma}\delta\gamma \text{ "=" } \frac{j(\gamma + \epsilon\delta\gamma) - j(\gamma)}{\epsilon}$$

for some "small" ϵ.

FLOW SOLVER

Compressible flows

Compressible lifting potential flows around 2D airfoils have been considered. We denote by Ω a domain surrounding the considered airfoil. To treat such flows, we consider that the wake Σ includes two separate lines Σ^+ and Σ^- joining the upper and lower surface (respectively) of the airfoil with the far field boundary. So, a new domain is defined, named Ω', which has the additional boundary part defined by Σ^+ and Σ^-. In this way, the potential jump is taken into account. The boundary condition is that the same flux has to cross Σ^+ and Σ^-, for treating Joukowski's condition. Since the flow is supposed to be isentropic, irrotational and uniform at the upstream boundary, the velocity $\mathbf{u}$ derives from a potential ϕ ($\mathbf{u} = \underline{\nabla}\phi$) satisfying the following unsteady equation.

Find the flow potential ϕ solution of

$$\frac{\partial \rho(\phi)}{\partial t} + \underline{\nabla}.(\rho(\phi)\underline{\nabla}\phi) = 0 \text{ in } \Omega' , \tag{1}$$

$$\rho \mathbf{u} \cdot \mathbf{n} \,|_{\Gamma_c} = 0 \; , \; \rho \mathbf{u} \cdot \mathbf{n} \,|_{\Gamma_{O\infty}} = \rho \mathbf{v}_\infty \cdot \mathbf{n} \; , \tag{2}$$

$$\rho \mathbf{u} \cdot \mathbf{n} \,|_{\Sigma^+} = \rho \mathbf{u} \cdot \mathbf{n} \,|_{\Sigma^-} \; , \tag{3}$$

$$p \,|_{\Sigma^+} = p \,|_{\Sigma^-} \; . \tag{4}$$

$$\phi = 0 \text{ on } \Gamma_{I\infty} \tag{5}$$

where ρ denotes the density and p the pressure given by the isentropic relation. The potential ϕ_∞ appearing in the Dirichlet condition (5) is such that $\mathbf{u}_\infty = \underline{\nabla}\phi_\infty$

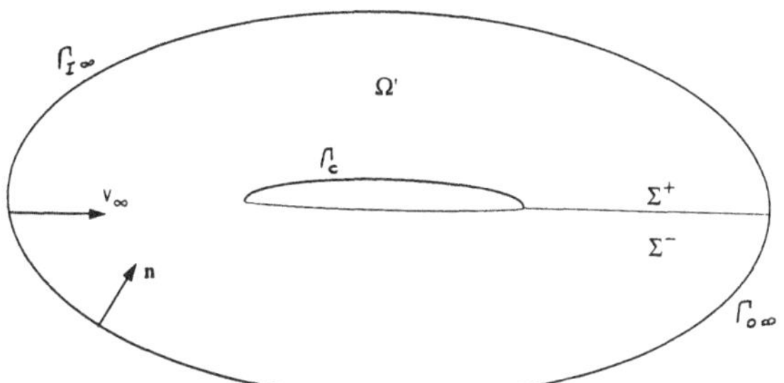

Fig.1: Computational domain around an airfoil

We denote by $\Gamma_{I\infty}$ the upstream far field boundary and by $\Gamma_{O\infty}$ the downstream far field boundary. We will denote by c the sound speed and by $M = \dfrac{|\mathbf{u}|}{c}$ the Mach number.

After having adimensionalized the equations, the density is then given by the relation:

$$\rho = (1 + \frac{\gamma - 1}{2} M_\infty^2 (1 - (2\frac{\partial \phi}{\partial t} + |\underline{\nabla}\phi|^2)))^{\frac{1}{\gamma-1}} \tag{6}$$

In order to treat transonic flows, one needs to introduce an upwind approximation active in supersonic regions. This upwinding is performed through the scheme of Engquist-Osher modifying the equation in the following way:

$$\left\{ \begin{array}{l} \underline{\nabla}_{upw}.(\rho \mathbf{u}) = \underline{\nabla}.(\rho \mathbf{u}) - \dfrac{\partial}{\partial \mathbf{s}}(\Delta s. \dfrac{\partial F}{\partial \mathbf{s}}) \\ \mathbf{s} = \dfrac{\mathbf{u}}{|\mathbf{u}|} \\ F = sign(M-1)^+ \; (\rho|\mathbf{u}| - \rho^*|\mathbf{u}^*|) \end{array} \right. \tag{7}$$

where Δs denotes the length of the considered element in the direction $\mathbf{s}$ and $()^*$ the sonic quantities.

The corresponding unsteady equations system for transonic potential flow become

$$\frac{\partial \rho}{\partial t} + \nabla \cdot (\rho \mathbf{u}) - \frac{\partial}{\partial \mathbf{s}}(\Delta s \frac{\partial F}{\partial \mathbf{s}}) = 0 \text{ in } \Omega' , \tag{8}$$

plus equations (2) – (5).

(9)

The solution of (8) with (2) - (5) such that $\frac{\partial \rho}{\partial t} \to 0$ when $t \to \infty$ is taken to be the solution of (1) - (5).

As time integration, a non-linear implicit formulation is considered and is given by the following time-stepping scheme:

$$\left\{ \begin{array}{l} \text{Find } \Delta\phi^n = \phi^{n+1} - \phi^n \text{ such that} \\ \frac{\rho^{n+1} - \rho^n}{\Delta t^n} + \underline{\nabla}.(\rho^{n+1}(\mathbf{u}^n + \underline{\nabla}(\Delta\phi^n))) = 0 \\ \rho^{n+1} = (1 + \frac{\gamma - 1}{2} M_\infty^2 (1 - (2\frac{\Delta\phi^n}{\Delta t^n} + |\mathbf{u}^n + \underline{\nabla}(\Delta\phi^n)|^2)))^{\frac{1}{\gamma-1}} \end{array} \right. \tag{10}$$

Weak formulation of the state equation

Thus, a variational formulation of equation (8) with related boundary conditions (2) - (5) is the following

Find $\phi \in V_0(\Omega')$ such that $\forall w \in V_0(\Omega')$

$$\int_{\Omega'} \frac{\partial \rho}{\partial t} w dx - \int_{\Omega'} \rho \, \mathbf{u} \cdot \underline{\nabla} w dx + \int_{\partial\Omega'_O} \rho \, \mathbf{u} \cdot \mathbf{n} \, w d\sigma - \int_{\Omega'} \mathbf{s} \cdot \underline{\nabla}(\Delta s \frac{\partial \mathrm{F}}{\partial \mathbf{s}}) \, w dx = 0, \tag{11}$$

with

$$V_0(\Omega') = \{w \in H^1(\Omega') : w \,|_{\Gamma_{I\infty}} = 0\}$$

and

$$\partial\Omega'_O = \Gamma_{O\infty} \bigcup \Sigma^+ \bigcup \Sigma^-.$$

Discretization

The potential is approximated by a piecewise linear fonction and the equation is discretized through a Galerkin variational formulation. Farfield boundary conditions on Γ_∞ are imposed through the knowledge of the unperturbed uniform flowfield. On the solid boundaries denoted Γ_P, no-slip conditions are assumed. Kutta-Joukovsky conditions are imposed through a cutting line which is in the mesh. All these boundary conditions are computed by adapting Riemann problems to isentropic equations and retaining the first component as mass flux as detailed in [3].

The solution of eq. (11) requires a spatial and time discretization. A P^1 finite element spatial discretization is used, therefore the discretized problem is (subscript h stands for discretized quantities and boundaries):

Find $\phi_h \in V_{0h}(\Omega'_h)$ such that $\forall w_{h_i} \in V_{0h}(\Omega'_h)$

$$\begin{array}{l} \int_{\Omega'_h} \frac{\partial \rho_h}{\partial t} \, w_{h_i} dx - \int_{\Omega'_h} \rho_h \, \mathbf{u}_h \cdot \underline{\nabla} w_{h_i} dx \\ + \int_{\partial\Omega'_{hO}} \rho_h \, \mathbf{u}_h \cdot \mathbf{n} \, w_{h_i} d\sigma - \int_{\Omega'_h} \mathbf{s} \cdot \underline{\nabla}(\Delta s \frac{\partial \mathrm{F}}{\partial \mathbf{s}}) \, w_{h_i} dx = 0, \end{array} \tag{12}$$

with

$$V_{0h} = \{w_h \in C^0(\Omega'_h) : w_h \,|_{T_i} \in P^1 \text{ and } w_h \,|_{\Gamma_{hI\infty}} = 0\}.$$

The numerical computation of the integrals on $\partial\Omega'_{hO}$ is performed by an adapted Riemann solver to isentropic equations and retaining the first component as mass flux (see details in [3]). The Osher's numerical flux has been chosen. The states on both sides of boundary are:

a) on $\Gamma_{hO\infty}$ the known farfield state g_∞ and the state obtained by the solution of the problem

b) on $\Sigma_h^+ \cup \Sigma_h^-$ the states obtained by the solution of the problem.

The discrete integration of the upwinding term is too technical to detail. It is proved that the value of this integral on an element T_l is

$$-(\mathbf{s}_i \cdot \underline{\nabla} w_{hi})\, F_h |T_l|,$$

where $\mathbf{s}_i$ is the unit vector on the velocity direction issued from point i, $|T_l|$ the measure of T_l .

The time discretization is done by the use of an implicit Euler scheme. A time step δt is chosen and supposing that k time steps have already been performed, we determine the quantities of next step by

$$\int_{\Omega'_h} \frac{\rho_h^{n+1} - \rho_h^n}{\Delta t^n}\, w_{hi} dx - \int_{\Omega'_h} \rho_h^{n+1}\, \mathbf{u}_h^{n+1} \cdot \underline{\nabla} w_{hi} dx \tag{13}$$
$$+ \int_{\partial\Omega'_{Oh}} \rho_h^{n+1}\, \mathbf{u}_h^{n+1} \cdot \mathbf{n}\, w_{hi} d\sigma + \int_{\Omega'_h} (\Delta s \frac{\partial F_h^{n+1}}{\partial \mathbf{s}}) \frac{\partial w_{hi}}{\partial \mathbf{s}} dx = 0, \quad \forall w_{hi} \in V_{0h}(\Omega').$$

By the help of equation (6) we can express $\rho_h^{n+1} = \rho_h^{n+1}(\frac{\phi_h^{n+1} - \phi_h^n}{\Delta t^n}, |\mathbf{u}_h^{n+1}|^2)$ with $\mathbf{u}_h^{n+1} = \underline{\nabla}\phi_h^{n+1}$. Equation (13) will give then $\delta\phi_h^{n+1} = \phi_h^{n+1} - \phi_h^n$, so the potential interferes as a computational trick with no physical meaning. The expression we consider for it is

$$\phi_h^n = \phi_{h\infty} + \phi_h^0 + \sum_{m=1}^{n-1} \delta\phi_h^m$$

where ϕ_h^0 stands for the initial potential.

Equation (13) is solved at each time step by non linear GMRES solver preconditionned by the incomplete decomposition of the Jacobian of the functionnal. The initialization of non linear GMRES is the solution of the linearized equation of eq. (13). This last solution is obtained by linear GMRES.

OPTIMIZATION PROBLEMS

One important class of problems to be addressed are the ones where the designer has some idea about the distribution that will lead to the desired performance. We consider here problems where the objective is to calculate the profile corresponding to a pressure distribution p^{obj} given in a certain number of points of the profile. We have to look for the minimization - zero for a reconstruction problem - of the function

$$j(\Gamma_P) = \int_{\Gamma_P} (p - p^{obj})^2 d\sigma \tag{14}$$

or

$$j(\Gamma_P) = \sum_{i=1}^{N} (p_i - p_i^{obj})^2 \tag{15}$$

where N is the number of chosen pressure points of the problem. In the following, the leading edge and trailing edge of the profile will be considered as fixed, so only $N-2$ control values will be actually taken into account.

Another kind of problems is to optimize some criterion related to aerodynamics subject to other constraints. One important class of problems is the minimization of drag for compressible flow cases. The problem to be solved is then given as :

$$\text{minimize } j(\Gamma_P) = \text{drag} \tag{16}$$

for a set **P** of admissible profiles

$$\mathbf{P} = \left\{ \begin{array}{l} \Gamma_P \text{ profile} \\ \text{its leading edge and trailing edge are fixed} \\ C_L \text{ is fixed} \end{array} \right\} \tag{17}$$

where C_L is the lift of the profile.

CONTROL VARIABLES

Different representations of the profile P have been considered :
1- Γ_P is defined by the y-coordinates of its mesh points while the x-coordinates remain fixed. If N denotes the number of profile mesh points, there are N-2 control variables.
2- Γ_P is defined by the y-coordinates of one over two mesh points while the x-coordinates remain fixed. It is a first step of inclusion of hierarchical strategy developped at INRIA Sophia-Antipolis
3- Γ_P is defined by two polynomial functions ($y = \mathcal{P}(x)$), one for the intrados and the other one for the extrados. Here, the control variables are the coefficients of these functions (the x-coordinates are fixed again).
4- Γ_P is represented by a third degree B-spline function. The control values are the (x, y)-coordinates of the points $(P_i)_i$ defined by :

$$P(t) = \sum_{i=1}^{m} P_i B_{i,3}(t) \tag{18}$$

where $P(t)$ is the parametrization of the profile and $(B_{i,3})_i$ are a basis of B-spline functions. The P_i are the vertex of the convex polygonal envelop of the B-spline curve. The incorporation of the strategy developped at INRIA Rocquencourt has been performed. Once control points have been chosen including leading and trailing edges of the airfoil, the number of discretization points between two successive control points is chosen. Then, a C^1 curve, piecewise polynomial of degree 3 (between two control points) is built joining the control points and with a tangent imposed at each control point in the following manner: the tangent is taken to be parallel to the chord joining the two neighbouring control points. Thus, we impose four conditions by control interval so that each polynomial of degree 3 is completely determined.

Furthermore, in each representation, the bound constraints that have been constructed at INRIA by [6] have been successfully introduced in our methodology in order to avoid the crossing of points of upper and lower lines of the profile.

SOLUTION ALGORITHMS

Inverse problems as defined in the previous section are solved by gradient methods (namely conjugate gradient methods) and by GMRES algorithm with no gradient computation. We refer to [6] and [5] and for applications to [6]. These methods have been used and compared for profile inverse problems [7] and [8].

For optimization problems, a penalization method is applied. The needed unconstrained minimization algorithm is a descent method using an estimated gradient. Two methods are considered to choose descent directions : the Polak-Ribiere method and the Fletcher-Reeves method.
The description of this minimization algorithm has been given in the 12 month report.

REMESHING METHODS

Three remeshing approaches have been investigated:

1) Energy minimization method (see [7]) by solving a linear elliptic scalar equation.

2) Linear elasticity theory for mesh deformation where the scalar equation is replaced by an elasticity operator coupling ccordinates of the mesh points (see [8]).

3) Spring analogy method developped by INRIA Sophia Antipolis and reported in the 12 month report has been investigated and seems of great robustness.

The mesh deformation based on the elasticity equation (method 2)) has been implemented in the design methodology based on the solution of compressible flows.

NUMERICAL RESULTS

The contribution to the Workshop has been focussed on performing 2D inviscid test cases, namely test cases T4 and T6.

Inverse problems for compressible flows (test case T4)

This test case was dedicated to an inverse problem in transonic regime. The problem is to recover the pressure coefficient C_p distribution of Korn airfoil starting from the NACA64A410 airfoil at a Mach number of $M_\infty = .75$ and no angle of attack.

We have applied the above methodology to reshape the given profile by taking the corresponding pressure distribution as a target one. For all the computations, control parameters are taken as the y-coordinates of the mesh points of the profile (representation 1). An O-mesh of 1103 nodes and 2200 elements has been used with 46 nodes on the profile. The GMRES algorithm has been considered with a Krylov space dimension of 5.

Firstly, a subsonic case $M_\infty = .3$ has been considered to test our approach in a simpler case. Corresponding results are synthetized on Fig.2 and it can be noticed that the functional is divided by a factor of 10^2 in about 200 state equation evaluations corresponding to a CPU time of about 200 minutes on a sequential IBM820 computer. Then after this number of iterations, the decrease becomes very slow.

Then, the actual test transonic test case has been computed. Fig.3 and Fig.4 present the results obtained respectively after 500 and 800 state equation evaluations. In this case, the decrease of the cost functional keeps the same slope up to 800 iterations; a division of this cost by a factor of 10^2 is obtained in about 600 state equation evaluations taking this time about 600 minutes on the same computer. The number of needed iterations has been multiplied by 3 between the subsonic and the transonic case.

Preliminary result for Workshop test-case T6

The test case is defined as the minimization of the drag on a RAE2822 at a given lift, for a Mach number of .73 and an angle of attack of 2^0. We present on Fig.5 a zoom of the 3108 nodes mesh with 126 nodes on the profile. On Fig.7, we present the distribution of pressure on the profile obtained with the compressible flow solver described previously. The considered functional is based on the drag coefficient C_d penalized to take into account the lift constraint C_l^{giv}:

$$j(\Gamma_P) = C_d \; + \; \mu \, (C_l \; - \; C_l^{giv}).$$

Two descent methods have been investigated : a fixed step gradient and a conjugate gradient algorithm. The profile has been parametrized by cubic splines as described in representation 4. A satisfactory representation of the profile has been reached with 20 control points (see Fig.8 for a detail near the trailing edge). We did not succeed within the last weeks of our work to obtain a real decreasing of the cost function although different parameters for the gradient estimation have been tested. The penalisation coefficient has not been maybe correctly tuned.

This strategy has to be more precisely analyzed and revised since rather successful results have been obtained when using GMRES solver on the optimality equation with INRIA Rocquencourt (see 24 month INRIA report).

CONCLUSIONS

After this 24 month activity, we can give an actual status of the achieved work within this BRITE/EURAM project by DASSAULT AVIATION with respect to the work program.

Concerning subtasks 1.1 and 1.6, non linear and aimed to be fast optimizers have been investigated, implemented and compared including conjugate gradient and non linear GMRES algorithms. Optimization with compressible flows (drag reduction) has not been achieved and appeared to be more challenging than originally planned.

Subtask 2.3 was dedicated to the inclusion of Euler flow solvers in an optimum design algorithm. This task has been completely supported by INRIA Sophia Antipolis and their developments will be implemented in our methodology in a near future.

Mesh deformation techniques (for surface and domain) investigated in subtask 3.1 have been clearly achieved. Tools have been implemented in our strategy; the implementation of the spring analogy method of INRIA is in progress.

In subtask 3.3 concerned with general sensitivity analysis, a wide range of comparison of algorithms have been performed (on the simple incompressible solver) and sensitivity of the overall method with respect to the number of mesh points is in progress. The algorithms developped at UPC should be incorporated in our methodology in the future.

Subtasks 4.1/4.3 have been achieved concerning inverse profile problems while duct problems have been extensively studied by INRIA Sophia.

Due to the complexity of the treatment of compressible flows in optimum design problems that we had to face, we did not have sufficient time to treat riblets problems with Navier-Stokes equations and this study has been initialized at INRIA Rocquencourt and Tritech.

ACKNOWLEDGEMENTS

Within the Brite/Euram project EUROPT I, we have worked in close cooperation with A. Dervieux, O. Pironneau from INRIA and D. Joannas, J.W. He and A. Vossinis.

REFERENCES

[1] J. CEA. *Conception optimale ou identification de formes: calcul rapide de la dériv'ee directionnelle de la fonction coût.* Notes de cours à l'Université de Nice.

[2] F. ANGRAND. *Thèse 3eme cycle.* Université Paris VI.

[3] Q.V. DINH and J.W. HE. *A Cartesian Grid Finite Element Method for Potential Flows.* 2nd Symposium on High Performance Computing, October 7-9, 1991, Montpellier, France.

[4] P.N. BROWN and Y. SAAD. *Globally convergent techniques in nonlinear Newton-Krylov algorithms.* Technical Report 89-57, Research Institute for Advanced Computer Science, 1989.

[5] P.N. BROWN and Y. SAAD. *Hybrid Krylov methods for nonlinear systems of equations.* SIAM J. Sci. Stat. Comp., 27, 1990.

[6] O. PIRONNEAU and A. VOSSINIS. *Comparison of some optimization algorithms for optimum shape design in aerodynamics.* INRIA Report, 1991.

[7] D. JOANNAS, B. MANTEL, J. PERIAUX and B. STOUFFLET. *12 month report BRITE/EURAM Optimum Design.*

[8] Q.V. DINH, D. JOANNAS, J.W. HE, B. MANTEL, J. PERIAUX and B. STOUFFLET. *18 month report BRITE/EURAM Optimum Design.*

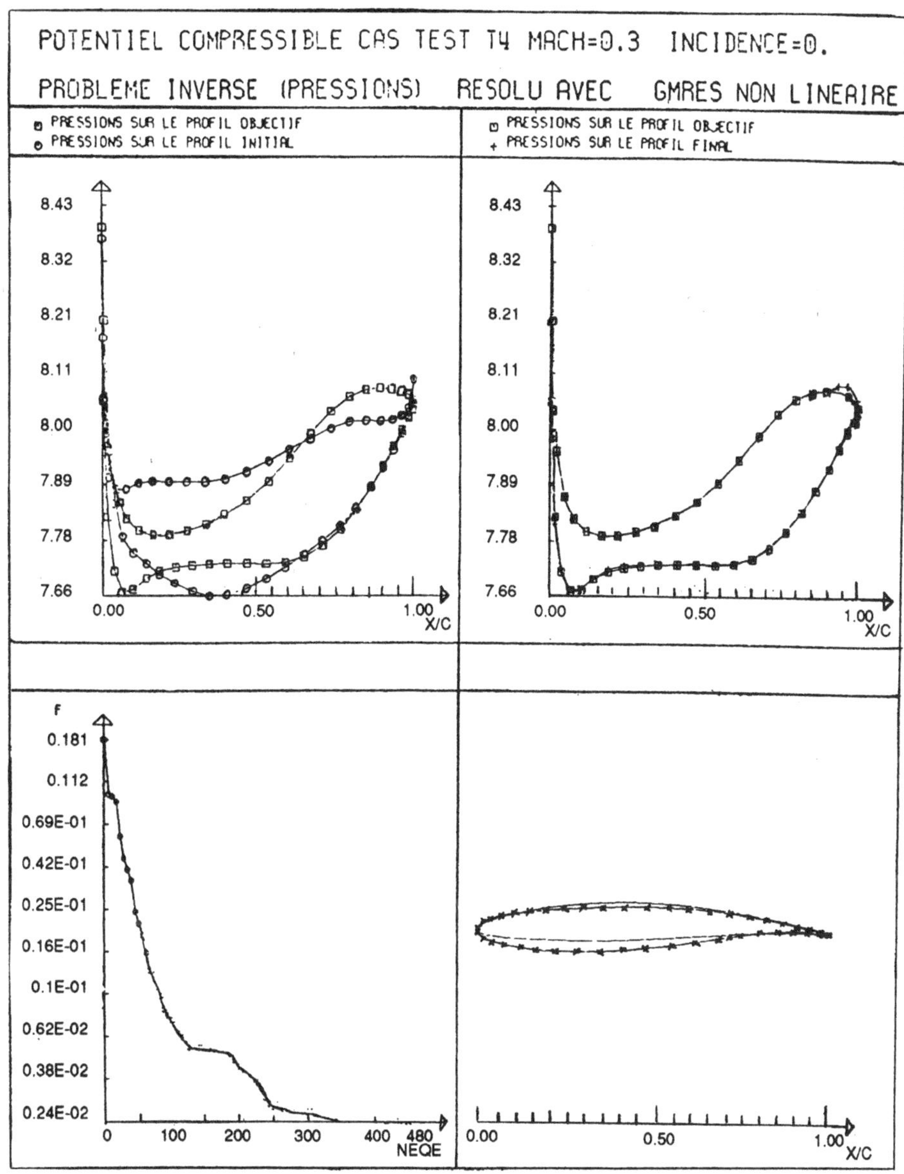

Fig.2: Preliminar results for test case T4 with $M_{\infty} = 0.3$

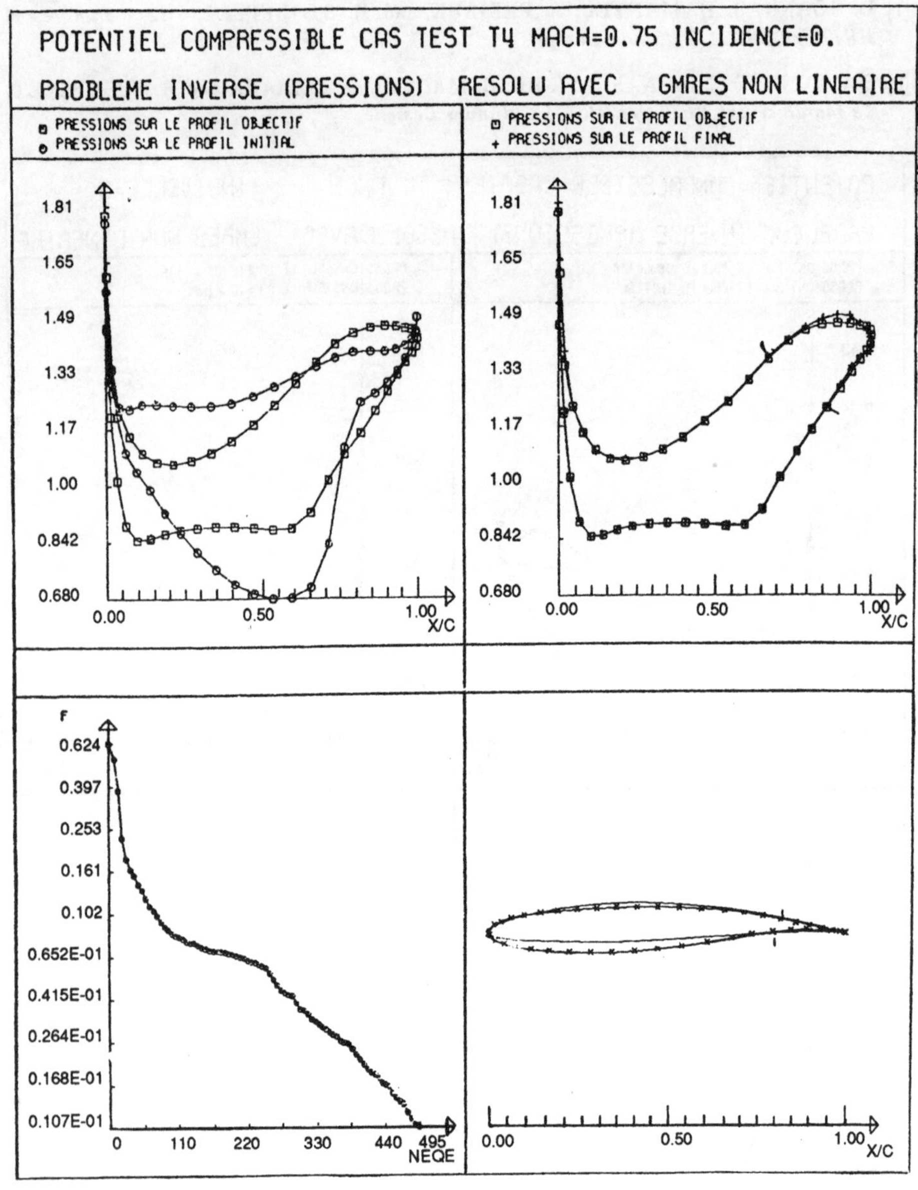

Fig.3: Results for test case T4 after 500 state equations evaluations.

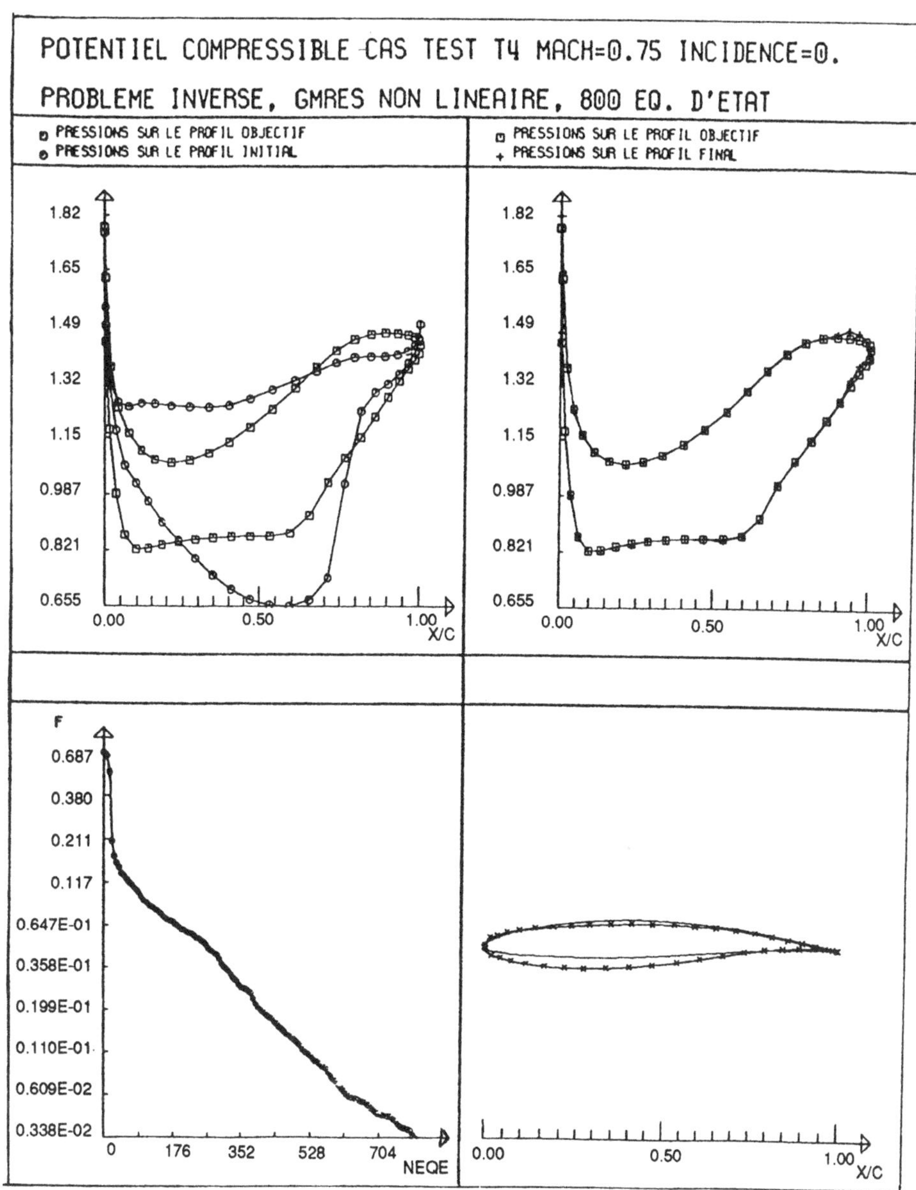

Fig.4: Results for test case T4 after 800 state equations evaluations.

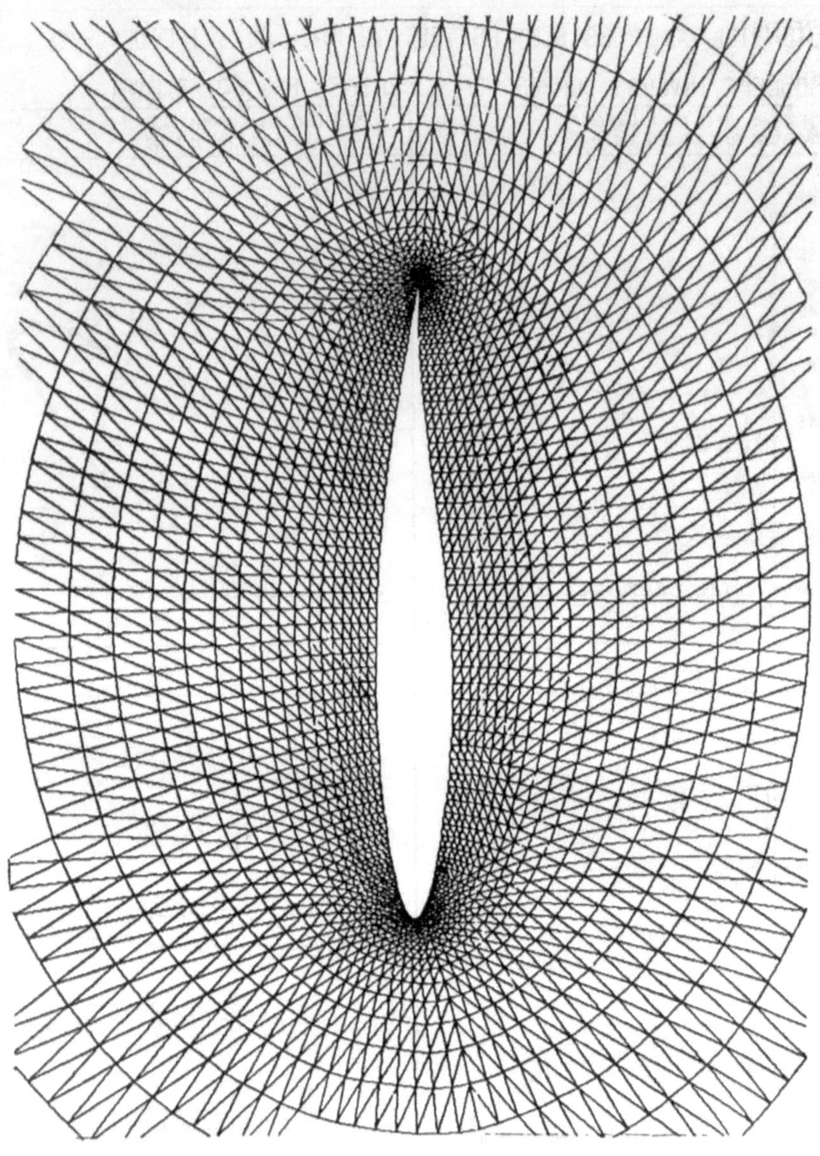

Fig.5: Finite Element mesh for the RAE2822 profile.

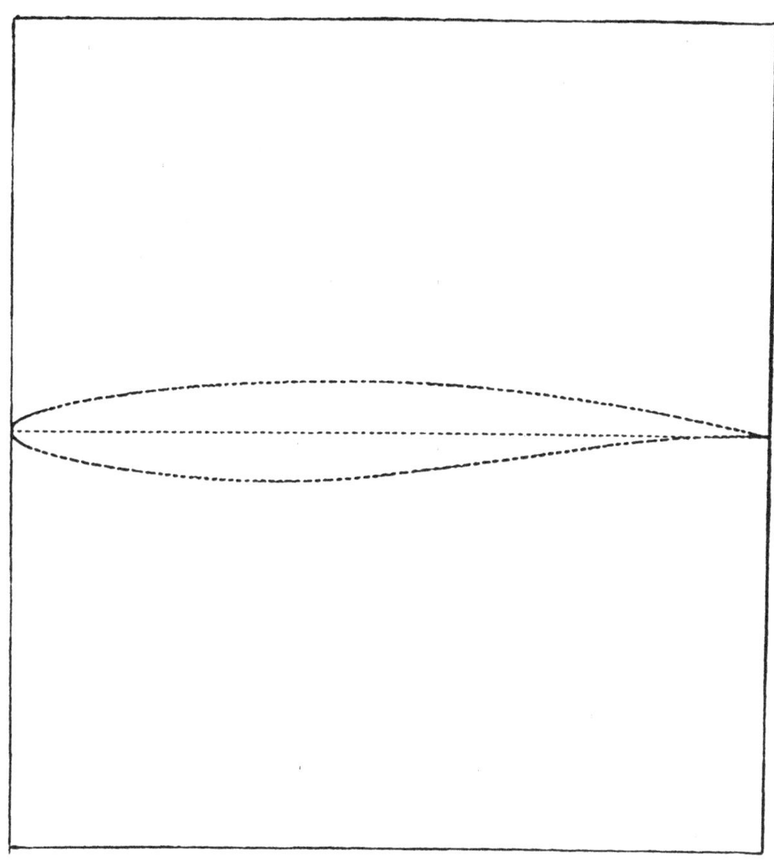

Fig.6: Representation of the RAE2822 profile.

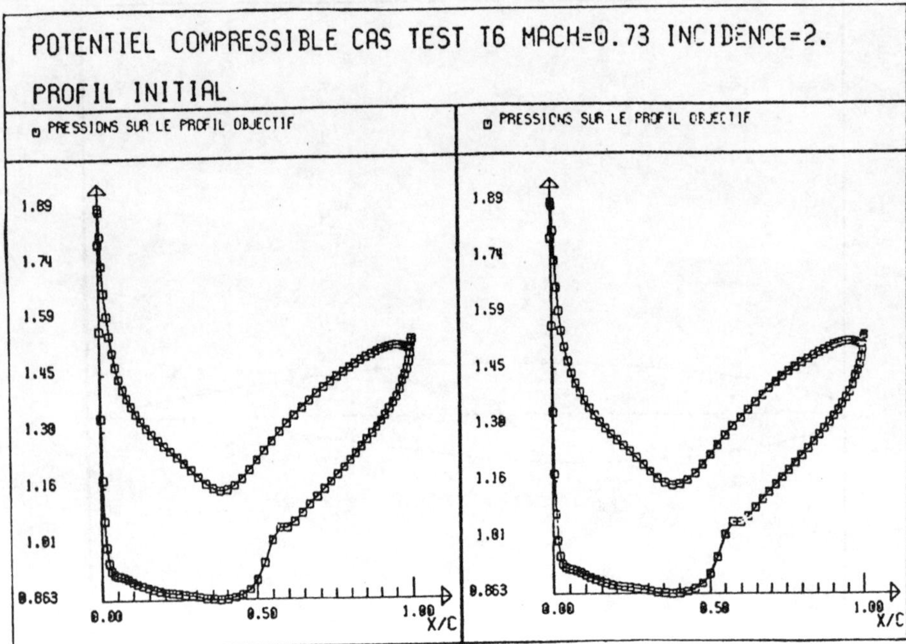

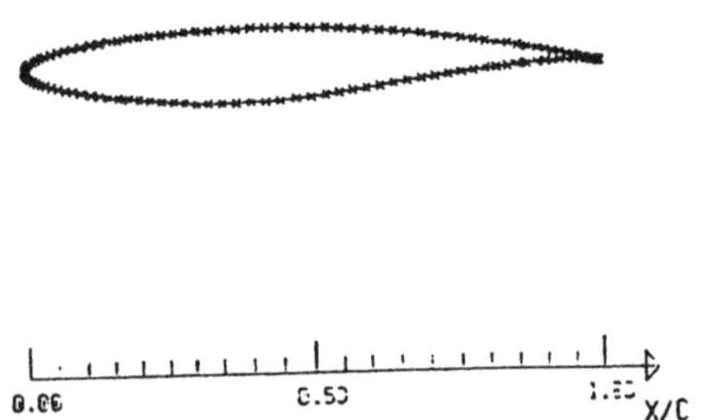

Fig.7: Pressure distributions on the RAE2822 profile.

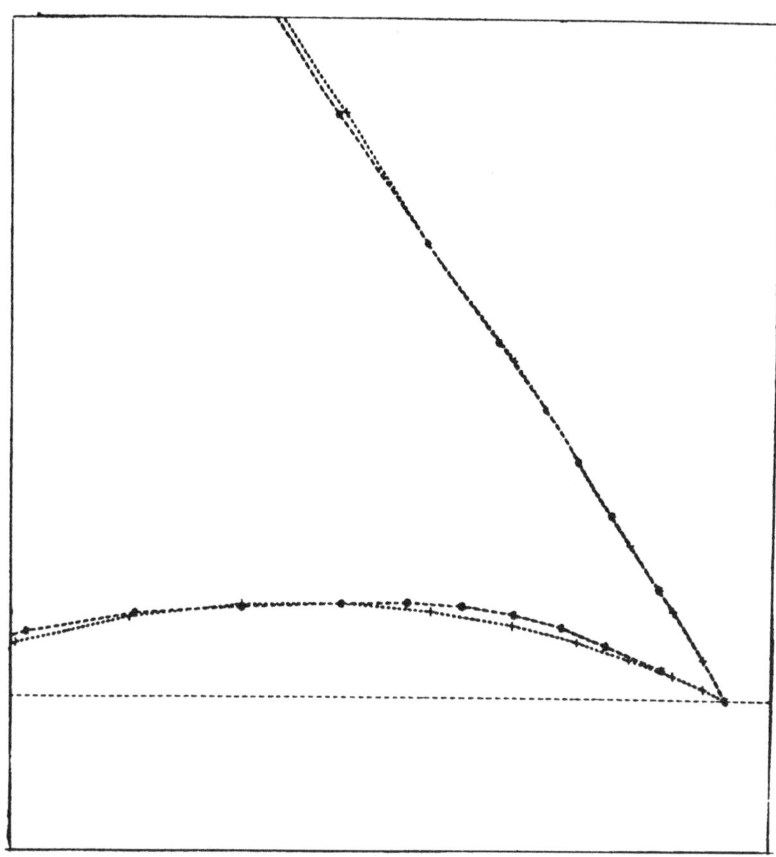

Fig.8: Detail of the control points for the representation of the trailing edge.

WING DESIGN WITH A 3D-SUBSONIC INVERSE PANEL METHOD

H. Schwarten

Deutsche Aerospace Airbus
Hünefelstraße 1-5, D-28199 Bremen, Germany

Abstract

A panel code, capable of treating complete configurations, has been extended by an invers option. This means, that two types of patches may be present: 'design patches', whose surface is derived from given pressures and 'analysis patches', whose pressures are derived from given surface. To solve such 'mixed' problems the panel code is embedded into an iteration loop, which minimizes the sum of squared pressure deviations. The geometry representation is done in terms of Bézier -splines with linear lofting. Intersections between wing- and body patches may occur. These are handled by locating a fictious design section inside the body and projecting it onto the body at a specified direction. Several examples demonstrate the power of the method and the usefulness to actual aircraft design problems.

INTRODUCTION

Nowadays, panel methods are an important investigation tool for the developement of modern aircrafts. In conjunction with some experimental reference data, subsonic characteristics of full aircraft configurations can be predicted with sufficient accuracy, see e. g. [1]. The usual way to use panel methods consists in specifiing the surface, or more exactly, a paneling of the surface, and then deriving pressure coefficients at the panel control points.

But, during aircraft development, the task arises, to meet specified cp-distributions at selected parts of the aircraft. In such situations we have to solve the inverse problem: derive the surface from given pressure distributions. To be more precisely, we treat the *mixed* problem, because at some parts of the configuration we solve for the surface (from given pressures), whereas at the remaining part we solve for the pressures (from given surface). This implies the two extremes of complete analysis and — in principle — complete design. However, a typical design task consists in reshaping a specific patch in order to design out the interference from some other parts of the airplane.

We are mainly interested in 3D-subsonic design, but the method works for 2D-cases, from which development started some times ago [2], too. Examples are presented for subsonic 3D-designs, transonic 2D-single element designs as well as subsonic 2D-multi element ones.

PRINCIPLE OF THE METHOD

The design task is formulated as a nonlinear, discrete minimization problem. The unknown part of the surface is described by a set of parameters $(a_1 \ldots a_n)$, called design parameters. These may be of very different type: angles, scale factors, expansion coefficients etc. They form a complete, unique description of the surface. After specifiing a discrete model, an analysis run of the underlying aerodynamic code would provide the pressure coefficients $\{c_{p_k}\}$ at a set of control points. Therefore, each pressure coefficient c_{p_k} is a given function of the design parameters

$$c_{p_k} = c_{p_k}(a_1 \ldots a_n) .$$

These functions should model a set of prescribed pressures $\{c_{p_k}^{target}\}$ in the least square sense. This means, that those values of the design parameters are to be determined, that minimize the sum of weighted, squared pressure deviations

$$E(a_1 \ldots a_n) = \sum_k \sigma_k \left(c_{p_k}(a_1 \ldots a_n) - c_{p_k}^{target}\right)^2 \stackrel{!}{=} min. \tag{1}$$

In the present approach of the invers task there are three main constituents: the aerodynamic code, the geometry description and the minimization procedure. These topics will now be discussed in detail, except from the first one for the following reason.

Here, the aerodynamic code is used in the 'analysis direction' only: surface in, pressure out. It could be regarded as a black box, providing pressure coefficients on demand. It needs not to possess any invers characteristics, so, in principle, every code may be used. However, to keep calculation time reasonable, some approximations should be possible, when calculating pressures changes due to surface perturbations. Normally, this can be done, but corresponding actions are code dependent (see later).

SECTION DEFINITION

A section is defined as the cut of the surface with a specific plane, mainly the coordinate planes. Sections are the fundamental elements for the built-up of both 2D- and 3D-geometries. Their location and shape has now to be expressed by a set of parameters in such a way, that most flexibility is gained.

Each section is defined relativ to a local frame, the section coordinate system SCS. The location, orientation and scaling of the SCS relativ to a global coordinate system GCS is specified by the so called global parameters. We take into consideration five such parameters $\vec{a}^{glob} = (t_x, t_y, t_z, \vartheta, l)$, where (t_x, t_y, t_z) locate the leading edge, ϑ sets the chord angle (rotation about y-axis) and l scales the chord. Depending on the surface patch to be designed, not all of these parameters may go free. They could be activated or deactivated by means of a control vector, supplied by input.

Contrary to the global parameters, local parameters specify the shape of a section relativ to the SCS. This is done in terms of Bézier curves, see Fig. 1. Let $\vec{r}(t)$ be the position vector of the curve relative to the SCS, then we have

$$\vec{r}(t) = \sum_{i=0}^{n} \vec{r}_i^{\,B} B_i^n(t), \quad 0 \le t \le 1 , \tag{2}$$

where $B_i^n(t)$ is the ith Bernstein polynomial of degree n

$$B_i^n(t) = \binom{n}{i} t^i (1-t)^{n-i}, \quad i = 0, \ldots n \tag{3}$$

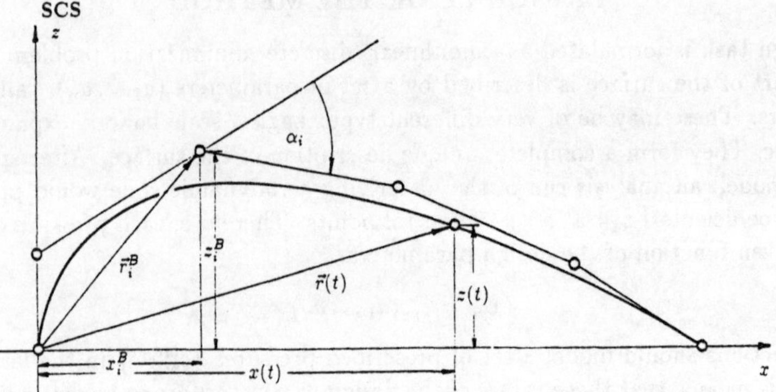

Figure 1: Bézier representation of a wing–type section.

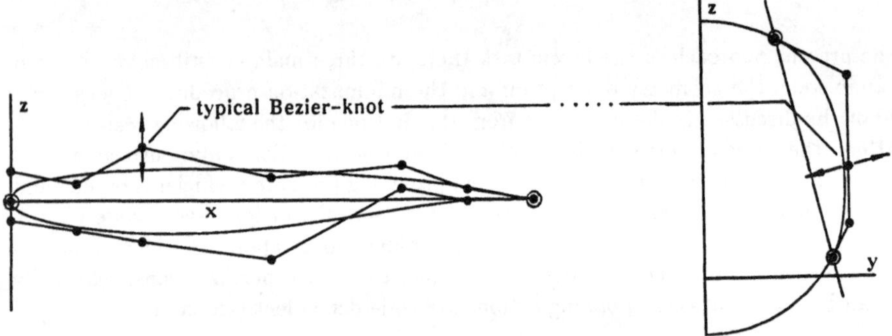

Figure 2: Allowed variations of Bézier knots.

and $\vec{r}_i^B$ is the position vector of the ith Bézier knot.

These curves are fundamental in geometric modeling, as is shown in standard text-books, for instance [3], [4], [5]. The Bézier knots (called control points as well) specify the shape of the curve. So they are possible local parameters, however, we don't use each of them for the following reason. Suppose e. g. a cut at y=const. (see left part of Fig. 2)

$$\vec{r}(t) = (x(t), z(t)) = \left(\sum_{i=0}^{n} x_i^B \, B_i^n(t), \sum_{i=0}^{n} z_i^B \, B_i^n(t) \right) .$$

The x-components x_i^B of the control points define the relation between the curve parameter t and the curve abscissa x. Variing the x_i^B (moving the control points in x-direction) leads to 'parameter optimization', i. e. finding the best curve parametrization. This may be critical (the Jacobian may become singular, see later), therefore control points are kept at fixed x–locations and are merely variated in z. In case of a generally located section control points are movable perpendicular to the section chord only (right part of Fig. 2).

We are left with the problem on how to distribute the control points along the chord. This is done as follows. The first and last one coincides with the first resp. last section point, causing the Bézier curve to pass through this points (for parameter values $t = 0$ resp. $t = 1$). The remaining control points are distributed nearly equidistant with a slight concentration towards the boundaries.

Closed sections ('wing-type') play an extra part. They are splitted into a lower and upper side, each with a separate Bézier representation. The normal tangent at the leading edge is obtained by placing the second control point above the first one, perpendicular to the chord. This reflects the fact, that the derivative of a Bézier curve at the endpoints is given by the direction of the adjacent control segment. Thus, both sides are coupled with C^1-continuity at the leading edge, which is sufficient for panel methods.

A very useful feature of Bézier curves is the *degree elevation property.* This means, that the same curve can be represented by control polygons with an increased number of points. A simple procedure relates both sets of control points (knot insertion algorithm). This allows a hierarchical design procedure, starting with few knots only. For successive steps with an increased number of knots, ideal initial parameter values can be constructed from the preceeding ones, i. e. such ones, that don't worsen the momentary value of the least square sum. The adaptability of the design sections is increased stepwise. This procedure has stabilized the minimium search significantly.

SURFACE DEFINITION

A patch is the smallest 'stand-alone' part of a configuration surface. It is built up from two or more sections, each given by a Bézier curve of degree n. To distinguish these curves, their control points now depend on a parameter λ

$$\vec{r}(t,\lambda)=\sum_{i=0}^{n}\vec{r}_i^B(\lambda)B_i^n(t), \quad 0\le t\le 1 .$$

Defining the dependence of the control points from λ to be a Bézier curve of degree m

$$\vec{r}_i^B(\lambda)=\sum_{j=0}^{m}\vec{r}_{ij}^B B_j^m(\lambda), \quad 0\le \lambda\le 1 ,$$

we arrive at the tensor product Bézier surface

$$\vec{r}(t,\lambda)=\sum_{i=0}^{n}\sum_{j=0}^{m}\vec{r}_{ij}^B B_i^n(t)B_j^m(\lambda), \quad 0\le t,\lambda\le 1 . \tag{4}$$

Most of the algorithms, valid for Bézier curves, could transfered to Bézier surfaces. But, to facilitate the *intersection* of patches, we restrict to linear dependence in λ-direction. Then, each couple of adjacent sections defines the surface part between them. Linear dependence means $m = 1$ and from (3) we get the two Bernstein polynomials

$$B_0^1(\lambda)=1-\lambda, \quad B_1^1(\lambda)=\lambda .$$

The resulting surface can be expressed by the equations

$$\begin{aligned}\vec{r}(t,\lambda) &= \vec{r}_0(t)(1-\lambda)+\vec{r}_1(t)\,\lambda \\ &= \vec{r}_0(t)+\lambda(\vec{r}_1(t)-\vec{r}_0(t)), \quad 0\le t,\lambda\le 1 ,\end{aligned} \tag{5}$$

with $\vec{r}_0(t)$ and $\vec{r}_1(t)$ as the defining sections

$$\vec{r}_0(t)=\sum_{i=0}^{n}\vec{r}_{i,0}^B\,B_i^n(t), \quad \vec{r}_1(t)=\sum_{i=0}^{n}\vec{r}_{i,1}^B\,B_i^n(t), \quad 0\le t\le 1 .$$

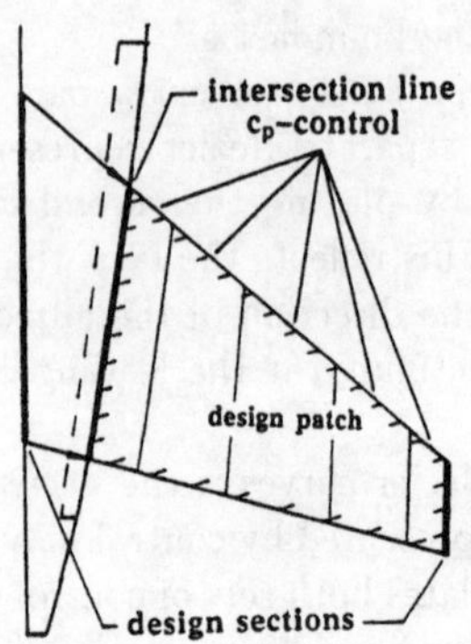

Figure 3: Definition of wing-body intersection.

The surface is piecewise *regular*, because it could thought to be obtained by continous movement of a straight line, connecting section points of equal t-parameter values, see (5). It is possible to keep selected sections fixed, thus maintaining a simple geometrical constraint in λ-direction.

INTERSECTION OF PATCHES

The definition of isolated surfaces in terms of tensor products is straightforward. Things become complicated, if intersections of patches occur. We are looking for an automatically executable procedure, because we want to design the intersection region itself. The method is illustrated by the wing-body intersection, see Fig. 3.

The wing is extended to the interior of the body by adding a *fictious* wing section at the symmetry plane $y = 0$. At each point of the inbody section a direction vector $\vec{s}^o$ is fastened, thus generating a generalized cylinder. This cylinder is then cut with the body surface giving the intersection line, which is used as *actual* inner wing section.

The fuselage surface is built up from sections at x=*const*. If we denote the left and right defining section by $\vec{r}_l(t) = (x_l, y_l(t), z_l(t))$ and $\vec{r}_r(t) = (x_r, y_r(t), z_r(t))$, the parameter λ is related to x by $\lambda = (x - x_l)/(x_r - x_l)$. Inserting this into (5), we get

$$\vec{r}_B(x,t) = \frac{x_r - x}{x_r - x_l}\vec{r}_l(t) + \frac{x - x_l}{x_r - x_l}\vec{r}_r(t), \quad x_l \le x \le x_r, \quad 0 \le t \le 1 .$$

The straight line, emanating from a point $(x_0, 0, z_0)$ of the fictious section, may be described by

$$\vec{r}_l(s) = (x_0, 0, z_0) + s\vec{s}^o, \quad 0 \le s < \infty .$$

Equating the expressions for the body surface and the line

$$\vec{r}_B(x,t) = \vec{r}_l(s)$$

yields three equations for the unknowns x, t and s. For a projection parallel to the y-axis the system decouples and essentially one nonlinear equation has to be solved.

The fictious wing section may be an ordinary design section of the wing, which means, that location and shape are determined from prescribed pressure distributions. Alternatively, it may be a fixed section. The design process treats fictious wing section in the same manner as the real ones; the only difference is, that they are replaced by

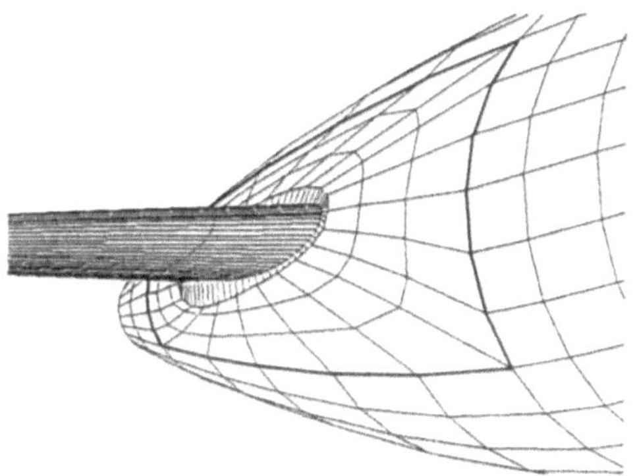

Figure 4: Wing-body intersection at rear airplane part.

the corresponding intersection lines before any c_p-calculations are done. The concept of using fictious, inbody sections is general enough, to apply to future configurations of interest, such as wing-pylon-nacelle intersections.

Fig. 4 shows the intersection between a horizontal tail and a rear fuselage with realistic shaped components.

MINIMIZATION PROCEDURE

The minimization of the sum of weighted, squared pressure residuals

$$E(\vec{a}) = \sum_k \sigma_k \left(c_{p_k}(\vec{a}) - c_{p_k}^{target}\right)^2$$

is done with a strategy due to LEVENBERG and MARQUARDT. It is a unification of two quite different procedures, the inverse Hessian method, which is good in the vicinity of the minimum, and the steepest descent method, which is good far from the minimum.

In the inverse Hessian method it is assumed, that the Taylor-expansion up to second order is a reasonable approximation of E in the vicinity of a current minimum location $\vec{a}_c$. A refinement $\vec{a}$ of this position is calculated in the following manner. The Taylor-expansion up to second order about the current location reads

$$E(\vec{a}_c + \vec{a}) \approx E(\vec{a}_c) + \sum_i \left.\frac{\partial E}{\partial a_i}\right|_{a_c} a_i + \frac{1}{2}\sum_{i,j} \left.\frac{\partial^2 E}{\partial a_i \partial a_j}\right|_{a_c} a_i a_j \,. \tag{6}$$

For abbreviation let

$$\beta_i \equiv -\frac{1}{2}\left.\frac{\partial E}{\partial a_i}\right|_{a_c} = -\sum_k \sigma_k \left(c_{p_k}(\vec{a}_c) - c_{p_k}^{target}\right) \left.\frac{\partial c_{p_k}(\vec{a})}{\partial a_i}\right|_{a_c} \tag{7}$$

be the gradient of E at the current location $\vec{a}_c$ and further

$$\alpha_{ij} \equiv \frac{1}{2}\left.\frac{\partial^2 E}{\partial a_i \partial a_j}\right|_{a_c} \simeq \sum_k \sigma_k \left(c_{p_k}(\vec{a}_c) - c_{p_k}^{target}\right) \left.\frac{\partial c_{p_k}(\vec{a})}{\partial a_i}\right|_{a_c} \left.\frac{\partial c_{p_k}(\vec{a})}{\partial a_j}\right|_{a_c} \,. \tag{8}$$

The matrix α_{ij} is known as Hessian matrix (apart from the factor 1/2). In the last formula we neglected the term proportional to the second derivative $\partial^2 c_{p_k}(\vec{a})/\partial a_i \partial a_j$. The Taylor-expansion (6) becomes

$$E(\vec{a}_c + \vec{a}) \approx E(\vec{a}_c) - 2\sum_i \beta_i a_i + \sum_{i,j} \alpha_{ij} a_i a_j \,.$$

From the condition for the minimum we get

$$\frac{\partial E}{\partial a_k} = 0 \implies -2\sum_i \beta_i \delta_{ik} + \sum_{i,j} \alpha_{ij}(\delta_{ik} a_j + a_i \delta_{jk}) = 0\,, \tag{9}$$

where the Kronecker symbol δ_{ij} stems from

$$\frac{\partial a_i}{\partial a_k} = \delta_{ik} = \begin{cases} 1 & i = k \\ 0 & \text{else} \end{cases}\,.$$

Carriing out the sums in (9), using the Kronecker symbol definition and the symmetry of the Hessian, we are left with

$$\sum_l \alpha_{kl}\, a_l = \beta_k\,. \tag{10}$$

The refinements $\vec{a}$, calculated from this equation, update the current minimum position $\vec{a}_c$ according to $\vec{a}_{new} = \vec{a}_c + \vec{a}$, which is the minimum position exactly, if the function E is just of second order.

For a steepest descent step, a step down the gradient, the formula for the refinement reads

$$a_k = const. \times \beta_k \tag{11}$$

with a suitably choosen constant, small enough not to exhaust the downhill direction. The last two equations (10) and (11) could be combined, as was pointed out by MARQUARDT. He showed, that the diagonal elements of the Hessian set the *scale* of the gradient. This is plausible from dimensionality considerations. E is nondimensional so β_k has the dimension of $[1/a_k]$, as may be seen from the definition (7). Therefore, the constant in the steepest descent formula (11) must have the dimension of $[a_k^2]$ i. e. is proportional to $1/\alpha_{kk}$. This can be expressed as a formula by introducing a factor λ (Marquardt–factor) and rewriting the steepest descent formula as

$$a_k = \frac{1}{\lambda \alpha_{kk}} \beta_k \qquad \text{or} \qquad \lambda \alpha_{kk} a_k = \beta_k\,.$$

Now both equations can be combined by defining a new matrix α' by the following prescription

$$\begin{aligned} \alpha'_{kk} &= \alpha_{kk}(1+\lambda) \\ \alpha'_{kl} &= \alpha_{kl}, \quad k \neq l \end{aligned}$$

and we are left with the single equation

$$\sum_l \alpha'_{kl}\, a_l = \beta_k\,. \tag{12}$$

If λ is large, the matrix α' is forced to be diagonally dominant and a steepest descent step is performed. On the other hand, as λ approches zero an inverse Hessian step is done.

The main idea of the LEVENBERG-MARQUARDT-minimization strategy is the adjustment of the factor λ during minimization. The calculation starts with a modest value of λ (e. g. 0.01) and is increased (more 'steepest descent'), if the least square sum worsens: $E(\vec{a}_c + \vec{a}) \geq E(\vec{a}_c)$, whereas λ is decreased (more 'Hessian'), if the least square sum diminishes. This method works very well in practise. Fig. 5 shows a flow chart of the main steps of the minimization strategy.

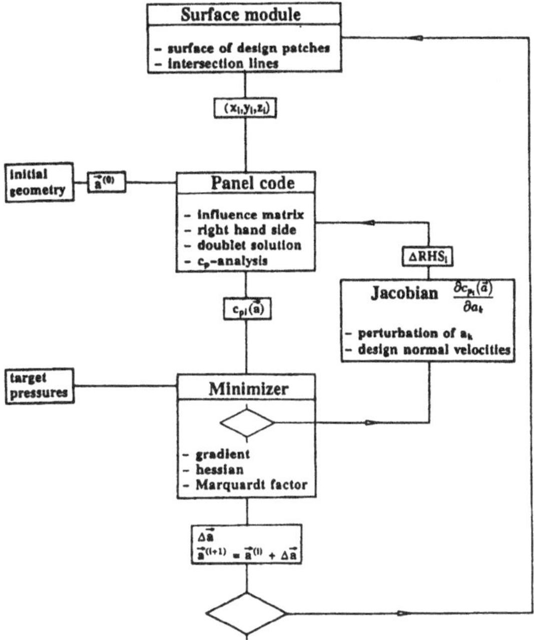

Figure 5: Flow chart of minimization strategy.

EQUIVALENT NORMAL VELOCITIES

The fundamental quantity, that must be known in performing minimization steps, is the matrix

$$J_{ik} = \frac{\partial c_{p_i}(\vec{a})}{\partial a_k} .$$

This matrix is the Jacobian of the transformation from design parameters to pressures. Their elements mean the change in pressure at control point i due to a variation of the design parameter a_k. The economical determination of the Jacobian is the crucial point of the present method. Here it is determined numerically — row by row — by performing analysis runs of the perturbated surface. But, and this is important, we *don't* have to perform a complete run through the panel code. Because small surface changes are involved we can use an 'equivalent-normal-velocity-concept', as was already done in [6]. This means, that we can take the influence matrix of the *unperturbed* surface, known from the preceeding iteration, along with a corrected right hand side in order to calculate the flow field about the perturbed surface.

Mathematically this means, that the LU-decomposition of the influence matrix hasn't to be done again. Therefore the CPU-amount for deriving the Jacobian is given by that of a back- and forward substitution, which is proportional to the square of panel number, whereas the amount for a LU-decomposition (a complete run) is proportional to the third power of panel number. Further, by using a special rank-one-update of the Hessian after a successful minimization step, several iterations can be done with the same Jacobian (three to four on average).

An expression for the equivalent normal velocities can be derived from the conservation of mass flux. Fig. 6 shows the unperturbed surface and the perturbed one, due to

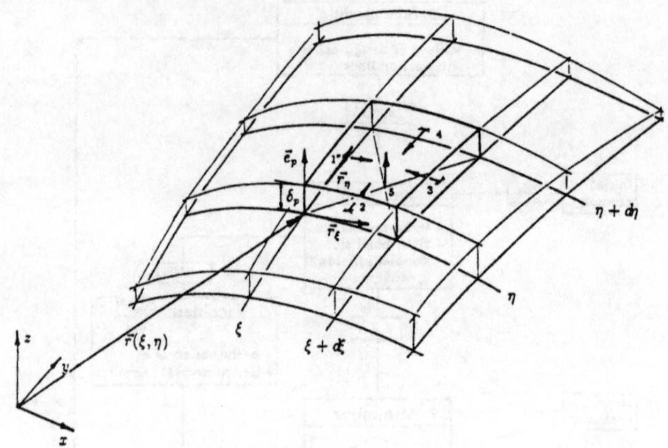

Figure 6: Mass conservation for infinitesimal polyhedron

the distortion of a certain Bézier knot in the direction $\vec{p}$. Conservation of mass flux for the infinitesimal polyhedron shown there means

$$\sum_{i=1}^{6}(\rho\vec{v}\,\vec{n}dF)_i = 0\,.$$

The boundary condition should be fulfilled on the perturbed geometry, so we have $(\rho\vec{v}\,\vec{n}dF)_6 = 0$. Contributions from opposite faces $(1 \leftrightarrow 3)$ and $(2 \leftrightarrow 4)$ can be combined and we can solve for the mass flux normal to the unperturbed surface. Inserting the appropriate surface elements — for example, the element of the unperturbated surface is given by $d\vec{F}_5 = [d\vec{r}_\xi, d\vec{r}_\eta] = [\vec{r}_\xi, \vec{r}_\eta]d\xi\, d\eta$ — yields the equation

$$\rho v_n\|[\vec{r}_\xi, \vec{r}_\eta]\| = \frac{\partial}{\partial\xi}(\rho\vec{v}[\vec{r}_\eta, \vec{e}_p]\,\delta_p) + \frac{\partial}{\partial\eta}(\rho\vec{v}[\vec{e}_p, \vec{r}_\xi]\,\delta_p)$$

ξ and η are the Gaußparameter of the surface given by $\vec{r}(\xi,\eta)$. $\vec{r}_\xi = \partial\vec{r}/\partial\xi$ and $\vec{r}_\eta = \partial\vec{r}/\partial\eta$ are tangent vectors to the parameter lines η = const. resp. ξ = const., $\vec{e}_p$ is a unit vector pointing in the distortion direction of the Bézier knot, δ_p is the resulting surface displacement (in that direction) and $\vec{v}$ is the velocity at $\vec{r}(\xi,\eta)$. [,] denotes the cross product of two vectors.

This method of calculating the Jacobian is restricted to panel methods. It is essential to 3D–designs, to make CPU–time tolerable, but is not so important for 2D–cases, for which the computational power of modern computers suffices, to do complete solutions even for parameter variations.

EXAMPLES

1.) Transonic 2D single–element design (Testcase T4)

The task is the reconstruction of the KORN airfoil at Machnumber $M_a = 0.75$. The used aerodynamic code is based on the full potential equation in body fitted coordinates, including a shock operator, satisfiing the RANKING–HUGONIOT relations [7]. Here, the Jacobian can't be calculated with the simplifications mentioned above, but we can do parameter variations on the same grid.

The profil is represented by eight movable Bézier knots for each the lower and upper profil half. The corresponding Bézier curve is of degree ten (the fixed control points at the leading and trailing edge have to be included).

Iteration starts with a NACA64A010, which is shown in Fig. 8a together with its Bézier approximation, c_p-distribution and target pressure distribution. Convergence is obtained after thirteen iterations. Fig. 8b shows the resulting resulting c_p-approximation and the Bézier representation of the designed airfoil.

Fig. 9 shows the convergence history, depending on 'number of elementary problems'. Here, an elementary problem is a call of the aerodynamic code. The horizontal parts of the curve then correspond to the calculation of the Jacobian. Each new Jacobian serves for three to four iterations. The relativ reduction of the L2-norm of the pressure deviations, available with each new Jacobian, diminishes with continuing minimum search. With the forth Jacobian the L2-norm is stationary.

Finally, Fig. 10 depicts the convergence history of the L2-norm of geometrical deviations between the designed an exact KORN-airfoil.

2.) Incompressible two-element design (Testcase T10)

The task is the reconstruction of WILLIAMS two-element configuration. The target pressures are potential theoretic exact, derived from a special conformal mapping [8].

The design is done using a panel code, based on linearly variing vortex distributions. Each profil half is controlled by 15 movable Bézier knots. The location of the leading edge of both components is known and fixed in space. Both elements are free to rotate around these points. Altogether, there are $2 \times (1 + 15 + 15) = 62$ parameters to fit 126 target pressures.

Iteration is started with two NACA0012 at zero incidence, see Fig. 11a. The converged solution together with the designed profiles is displayed in Fig. 11b. The target pressures are met very well, besides from small deviations near the trailing edge. This is not surprising, because the differences in aerodynamic modeling—conformal mapping versus panel method— are largest here. The convergence history is shown in Fig. 12 (compare the explaining remarks of example 1.). The Jacobian has been calculated four times, but from Fig. 12 it is seen, that actually three Jacobians, serving for nine to ten iterations, would have been sufficient.

A comparison between the designed and the original profils is shown in Fig. 13. A doubled z-scale is used to display differences properly. The chord angles of the design profiles are $\vartheta_1 = 1.67$ resp. $\vartheta_2 = 10.13$, whereas the corresponding original values amount to $\vartheta_1 = 1.51$ resp. $\vartheta_2 = 10.00$. The deviations in chord angles and shape stem from the different aerodynamic modeling (conformal mapping versus panel method). The c_p-distributions fit, but these intrinsic differences are translated to geometric ones by the inverse code!

3.) Subsonic 3D wing-body combination (Testcase T14)

The task is the reconstruction of a 3D-wing, attached to a body. Fig. 14 depicts the panel model consisting of 826 panels. The original wing is defined by eleven sections, giving a target pressure distribution with ten columns in span each with 48 pressure control points.

The designed wing will be built up from five design sections, each of which is controlled by six movable Bézier knots per half. Furthermore, each design section is free to rotate about the trailing edge. Altogether $65 = 5(1 + 6 + 6)$ parameters should fit the $10 \times 48 = 480$ target pressures.

The used aerodynamic code is the VSAERO–code [9]. It is a subsonic panel code, based on piecewise constant doublet and source singularities, capable of treating complete aircraft configurations. All following 3D–examples will use this code.

Fig. 15a and 15b show the comparison of the initial, target and converged pressure distributions at each of the ten spanwise stations. All target pressures are met well. In Fig. 16 the designed profils are compared with the original ones. Here, cut S3, S4 and S5 show some differences at the lower side. This belongs to the fact, that the designed wing is built up from only five, but the original one from eleven sections. Finally, Fig. 17 shows the L2–norm of the pressure deviations at five selected spanwise stations and the total L2–norm. The Jacobian is calculated twice, but essentially once would have been sufficient.

4.) Wing–nacelle interaction

The configuration consists of a simple 3D–wing and a nacelle, Fig. 18. The task is to eliminate the influence of the nacelle on the wing. This means, that the pressure distribution of the clean wing defines the target distribution for the interacting system. The wing is divided into two patches, from which the inner one is a design patch with four design sections S1, S2, S3, S4 and a fixed section, which establishes the coupling to the rigid outer wing patch. The configuration is symmetrical to the midplane of the nacelle.

Each design section is represented by thirteen parameters (chord angle, six variabel Bézier knots for each the lower and upper side) giving a total of fiftytwo parameter. These have to fit 208 prescribed pressures.

Fig. 19 shows the c_p–distributions at the four c_p–control cuts. The triangles represent the target pressures; the circles indicate the uncorrected situation (0. Iteration). The differences between them correspond to nacelle interference. The crosses mark the solution obtained after four iterations: the interference has been designed out.

The comparison of the initial wing sections (dashed line) with the designed ones (solid line) in the global reference system is shown in Fig. 20a. The interference affects both the chord angle and the shape. The designed profiles show increased incidence. To separate these effects, the profiles S1, S2 and S3 (S4 is left out, because the difference isn't displayable) are compared in normalized positions at an enlarged z–scale, see Fig. 20b. From this it is seen, that only the lower sides are reshaped. The designed profiles have increased thickness.

5.) Redesign of fuselage rear part

The last two examples dealt with the design of patches with target pressures specified on *themselves.* But this is a condition, that must not be fulfilled. In this last example we demonstrate a special feature of the present method, namely the splitting of target pressure patches from design patches.

In Fig. 21 the rear part of an airplane with fuselage–tail, fin, pylon and nacelle is depicted. The design task is to reduce the rather high velocity level ('channel effect') on the pylon. But, instead of reshaping the pylon, this task can be solved by appropriate changes of the fuselage in the pylon attachment region.

Fig. 22 displays the initial pressure distribution (circles), the target distribution (triangles) due to a reduced velocity level, as well as the converged pressure distribution (crosses).

Fig. 23 shows a comparison of the initial fuselage sections (solid lines) and the designed ones (dashed lines), which are represented by cubic Bézier curves. The waisting of the body sections is controlled by pressures on the pylon!

REFERENCES

[1] Stuke, H., *Analyse des MPC75 Windkanalversuchs WP9007/9006 Propfan-Konfiguration*, Deutsche Aerospace Airbus, Bremen, Bericht EF–1820, 1990.

[2] Schwarten, H. *Ein inverses, subsonisches 2D–Panelverfahren nach der Methode der kleinsten Quadrate zum Entwurf und zur Modifikation von Mehrelementprofilen*, DGLR–Jahrbuch 1987 (1) der Deutschen Gesellschaft für Luft–und Raumfahrt, pp. 163–170.

[3] Yamaguchi, F., *Curves and surfaces in computer aided geometric design*, Springer Verlag, Berlin, Heidelberg, New York, 1988.

[4] Farin, G., *Curves and surfaces for computer aided geometric design*, Academic Press, London, Sydney, Tokio, Toronto, 1990.

[5] Hoschek, J., Lasser, D. *Grundlagen der geometrischen Datenverarbeitung*, B. G. Teubner, Stuttgart, 1989.

[6] Kubrynski, K. *A subsonic panel method for design of 3–dimensional complex configurations with specified pressure distributions*, Notes on Numerical fluid mechanics, Vol 21, Vieweg Verlag, Braunschweig, Wiesbaden, 1988, Proceedings of the third GAMM–Seminar Kiel, January 16 to 18, 1987, pp. 137–146.

[7] Jakob, H., *Ein Verfahren zur Berechnung der ebenen transsonischen Strömung in Stromlinienkoordinaten*, ZKP–Ergebnisbericht Nr. 48, 1985.

[8] Williams, B. R., *An exact test case for the plane flow about two adjacent lifting airfoiles*, ARC R&M 3717, 1973.

[9] Maskew, B., *Prediction of subsonic aerodynamic characteristics: a case for low order panel methods*, Journal of Aircraft, Vol. 19, No. 2, February 1982, pp. 157–163.

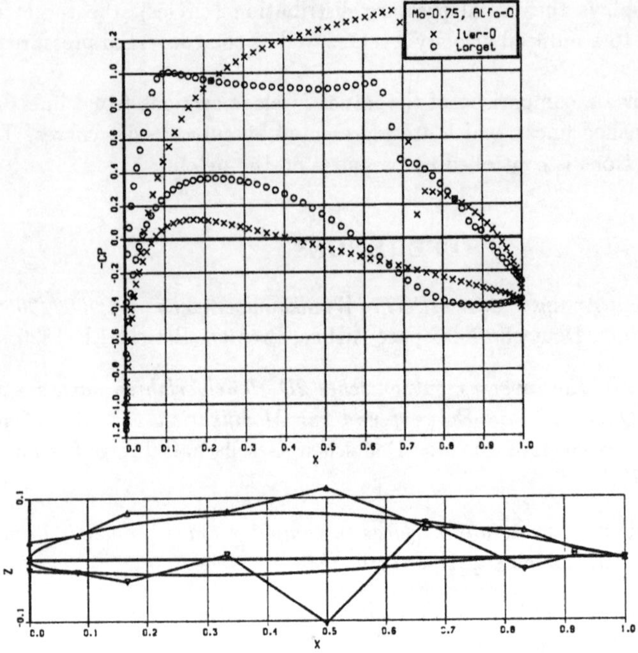

Fig. 8a Initial configuration for recovery of KORN–profil.

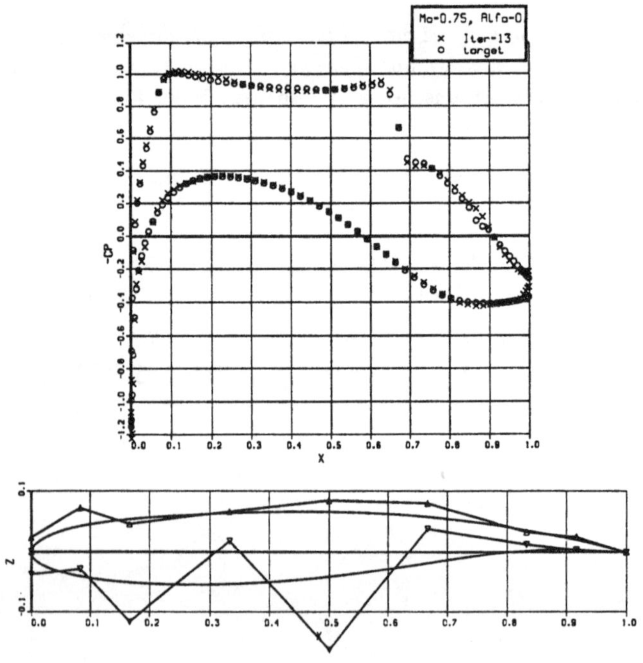

Fig. 8b Bézier –approximation of KORN–profil, M=0.75, Iter=13.

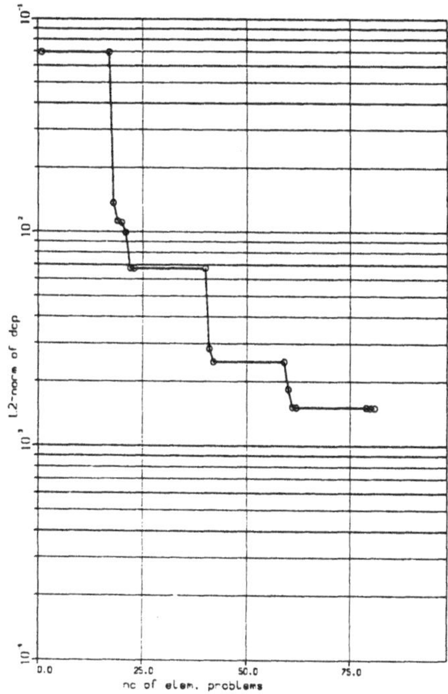

Fig. 9 Convergence history of L2–norm of c_p–residuals.

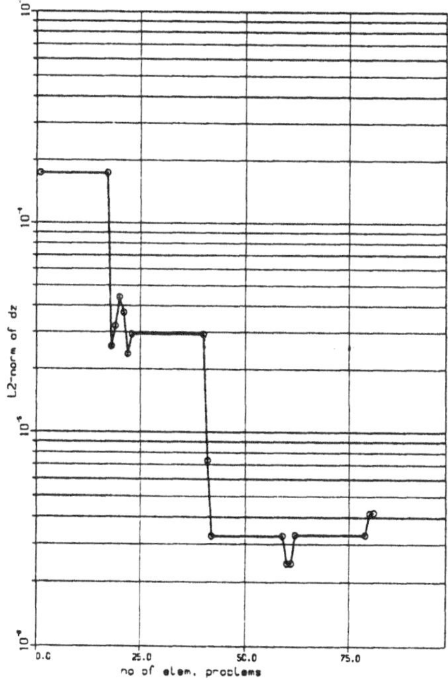

Fig. 10 Convergence history of L2–norm of shape residuals.

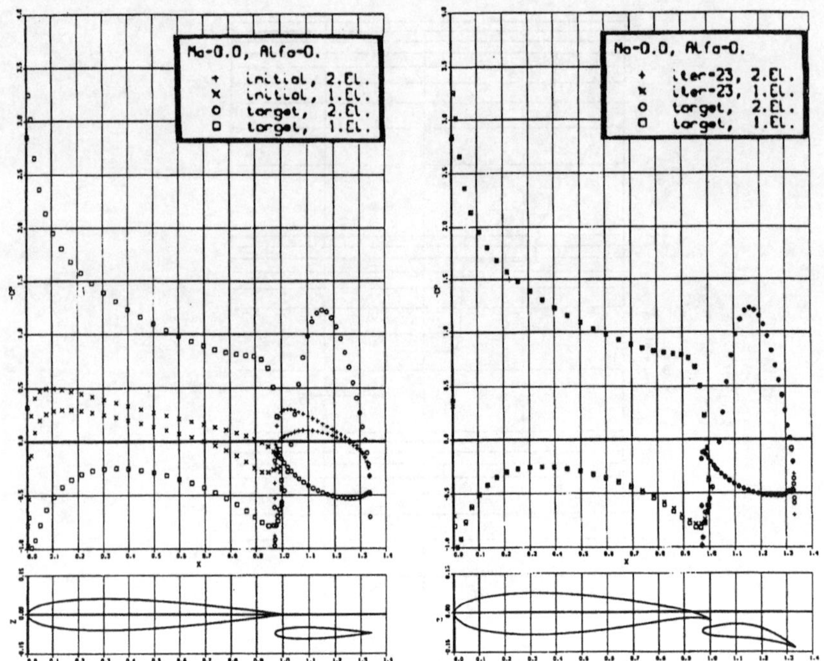

Fig. 11a Initial configuration for recovery of William's two–element case.

Fig. 11b Converged state.

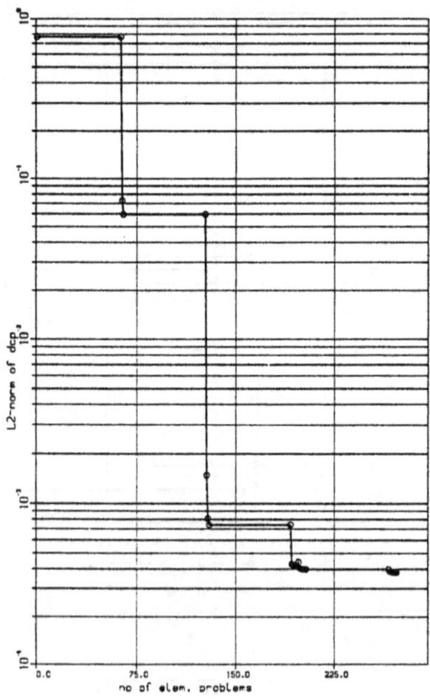

Fig. 12 Convergence history of L2–norm of c_p–residuals.

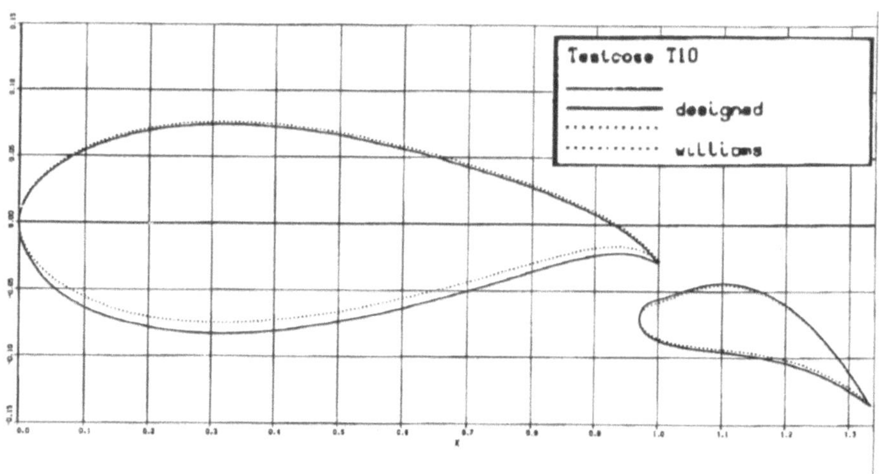

Fig. 13 Comparison of designed and original William's profiles (enlarged z-scale).

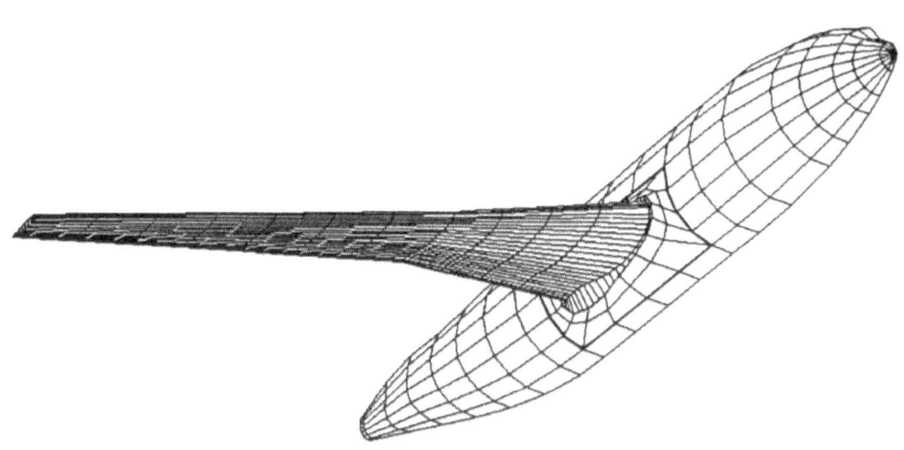

Fig. 14 Panel model for wing–body testcase.

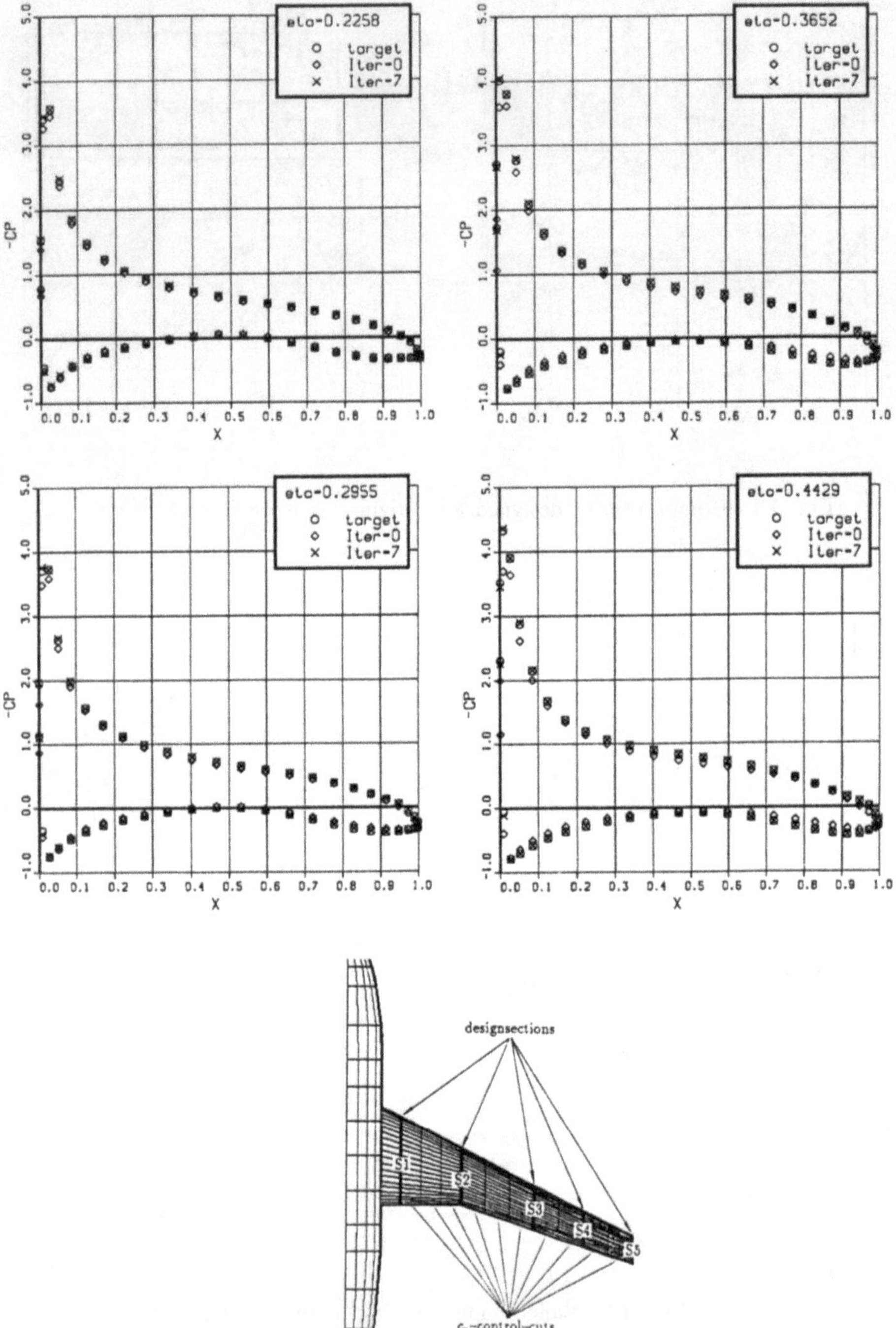

Fig. 15a Comparison of initial, target and converged pressure distributions at inner four stations.

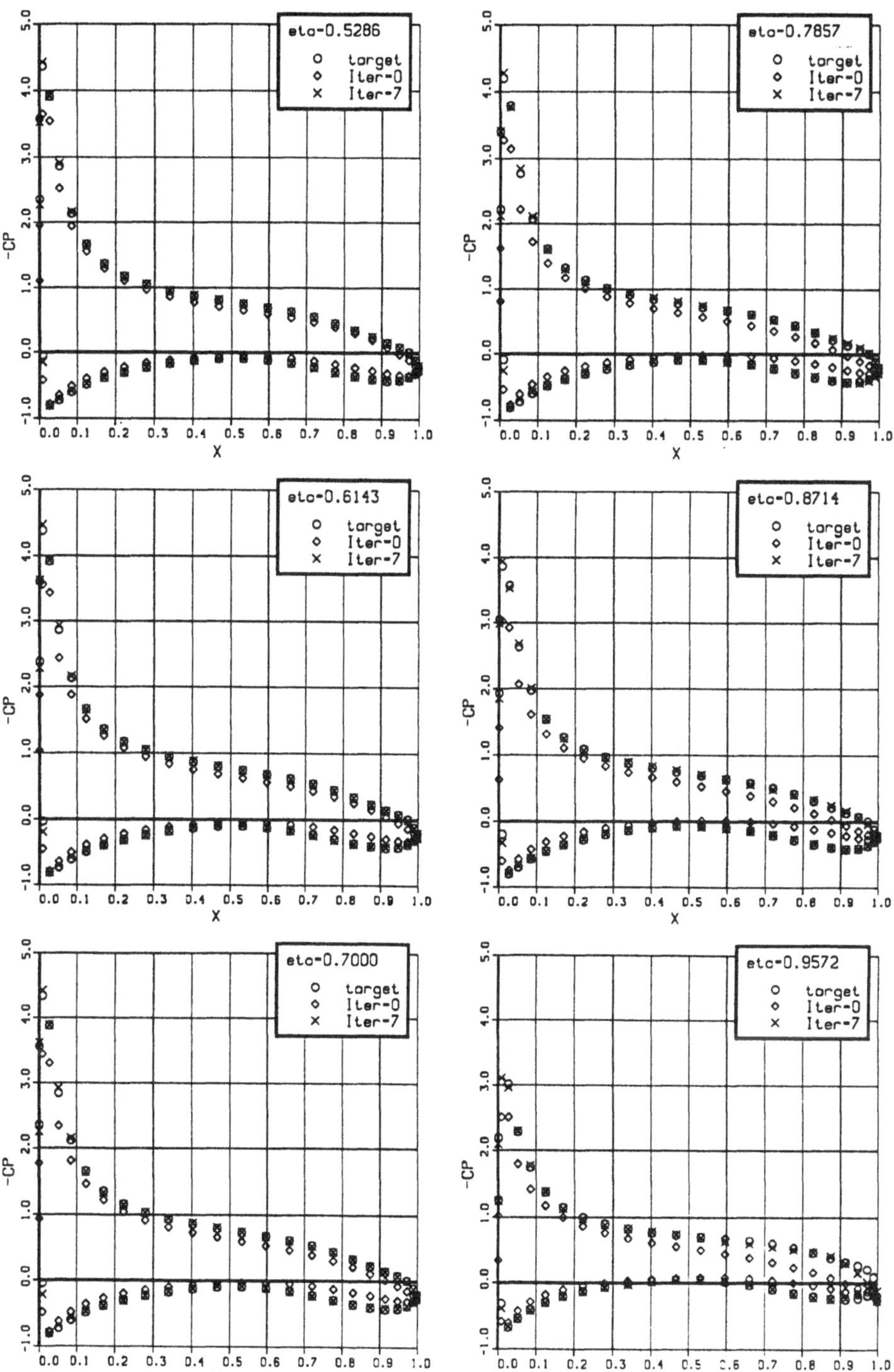

Fig. 15b Comparison of initial, target and converged pressure distributions at outer six stations.

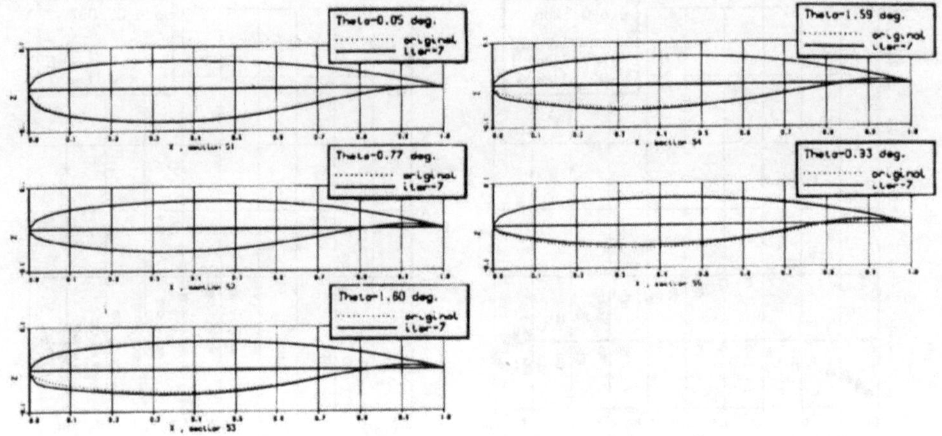

Fig. 16 Comparison of design sections with original ones (normalized position).

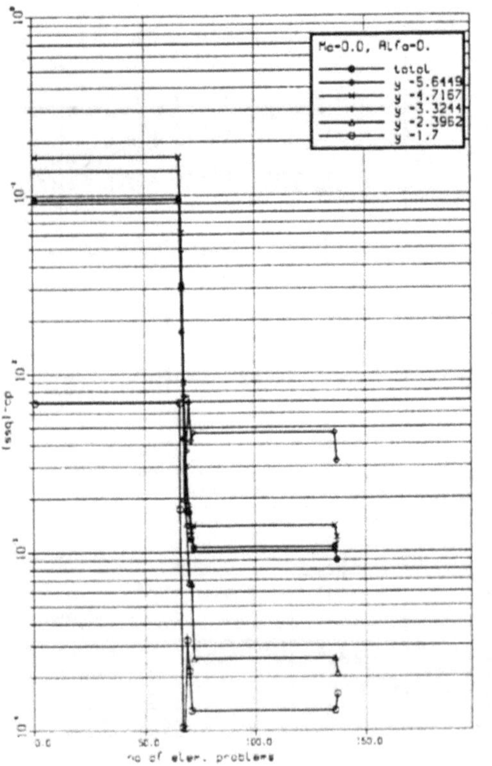

Fig. 17 Convergence history of L2–norm of c_p–residuals at five selected stations.

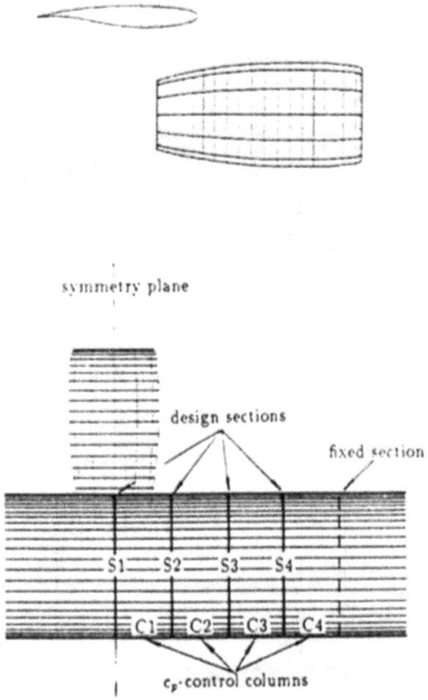

Fig. 18 Wing–nacelle assembly.

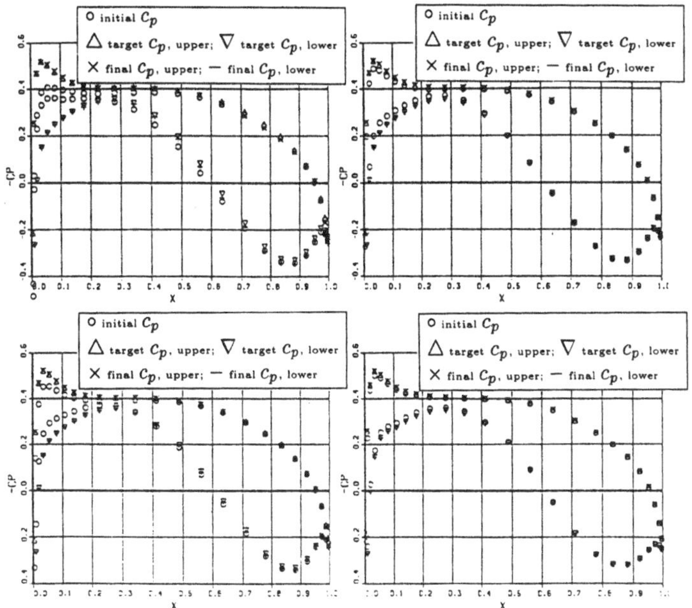

Fig. 19 Comparison of pressure distribution, $M_a = 0.$, $\alpha = 0$.

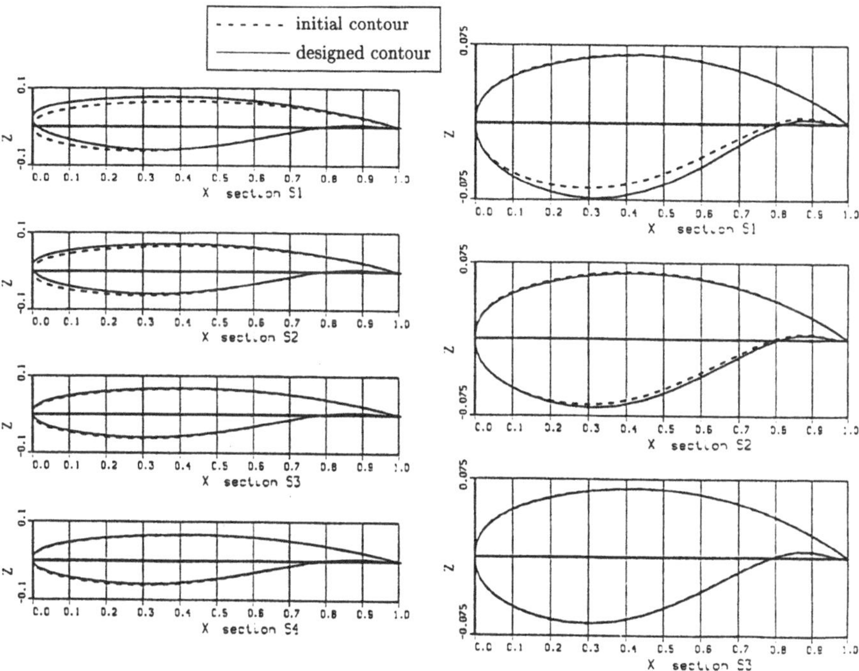

Fig. 20a Comparison of initial and designed profiles.

Fig. 20b Comparison of normalized profiles.

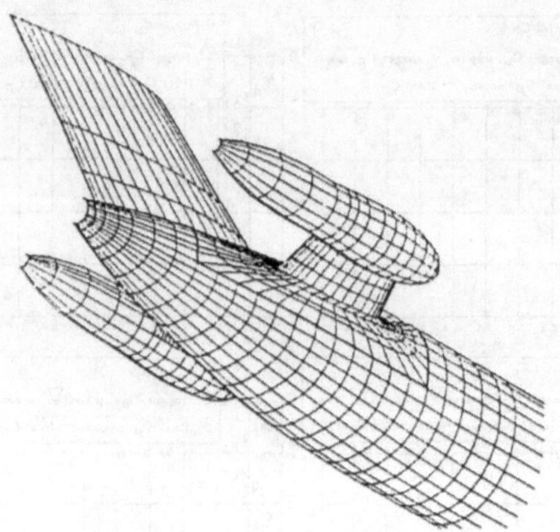

Fig. 21 Design of fuselage rear part.

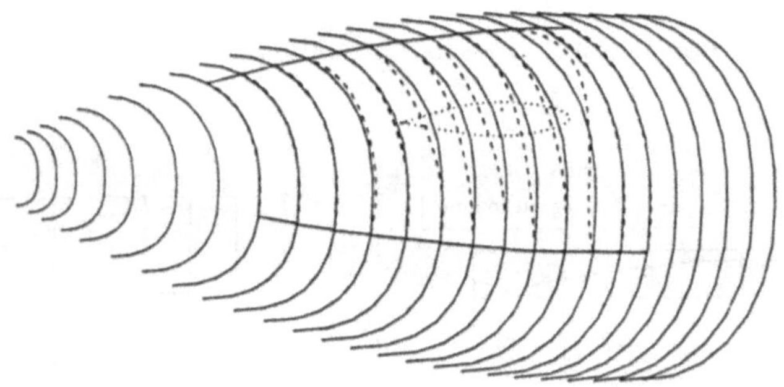

Fig. 22 Comparison of initial and designed body patch.

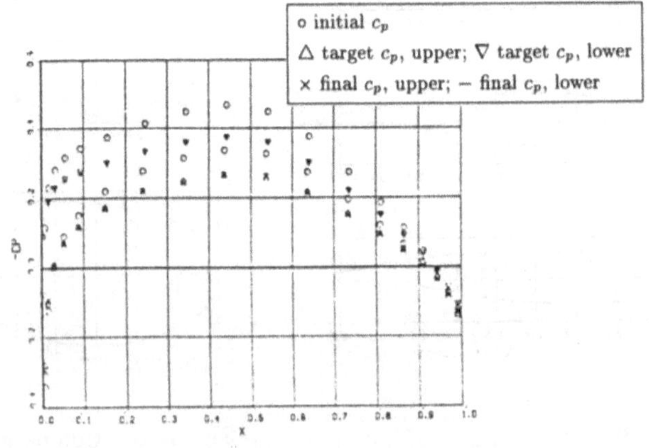

Fig. 23 Initial and designed Pressure distribution on inner pylon cut.

DIRECT OPTIMIZATION OF TRANSONIC AIRFOILS

V. Selmin
Alenia Aeronautica D.V.D., TEVT
Corso Marche 41 , I-10146 Torino

SUMMARY

The paper reports on the numerical procedure for the optimization of transonic airfoils developed in ALN D.V.D. within BRITE/EURAM AERO-0026 contract. Results for the solution of test case T6 of the workshop are presented.

INTRODUCTION

One of the most important and challenging problems in aircraft configuration optimization is the design of the shape of wings sections. Two different approaches to this problem can be identified in the literature on the subject.

The more classical approach involves the use of the so-called inverse methods. By means of inverse methods, wing section shapes that will generate a given pressure distribution at a given freestream Mach number can be found. The task of the aerodynamicist is to specify a pressure distribution such that the wing geometry that will be found performs aerodynamically as required.

A more recent and possibly a more promising one is called direct optimization. In this approach an aerodynamic object function, such as drag, is minimized computationally by varying the parameters that describe the geometry of the wing. While doing so, the solution may be subject to geometric and/or aerodynamic constraints. The main difficulty is in successfully choosing the succession of transitions in the airfoil shapes that transforms the starting airfoil into the final one which meets the stated goals (Fig. 1). This problem can be solved by using an optimum design procedure.

In general, an optimization procedure is composed by an optimization module, an aerodynamic module and an interface module that also handles the geometric modifications of the airfoil (Fig. 2). The aerodynamic analysis module usually is a well assessed CFD computer code whose reliability has a strong impact on the final result. On the other hand, the code must have a low computational cost in order to maintain the overall optimization cost at a reasonable level. However, highly sophisticated CFD computer codes can be used to verify the aerodynamic characteristics of the optimized airfoil.

ALN D.V.D. has developed such an optimum design procedure for the design of transonic airfoils. This paper reports on some numerical experiments we have performed to assess the reliability of our code.

OPTIMIZATION PROCEDURE FOR THE DESIGN OF AIRFOILS

To solve the optimization problem the values assigned to the design variables $\mathbf{X}$ must be found so as to minimize the objective function $F(\mathbf{X})$ while maintaining that the possible constraints functions $G_j(\mathbf{X})$ are ≤ 0. The designer must derive this functions and choose the design variables that govern the transformation of the airfoil geometry. Elements of these functions are the aerodynamic coefficients, which are obtained by a CFD analysis code.

Objective function and constraints

The general aerodynamic design requirements need to be translated in mathematical formulas for constraints and objective functions. A simple expression for the objective function includes the linear combination of aerodynamic coefficients (C_d, C_m, C_l, C_p) at one or more operating conditions of the airfoil (n). For instance,

$$F(\mathbf{X}) = \sum_{i=1}^{n} \{ \; K_{1,i}\, C_l(\alpha_i, M_{\infty i}, \mathbf{X}) + K_{2,i}\, C_d(\alpha_i, M_{\infty i}, \mathbf{X}) + K_{3,i}\, C_m(\alpha_i, M_{\infty i}, \mathbf{X}) + K_{4,i}\; \Phi\,(C_p(\alpha_i, M_{\infty i}, \mathbf{X})) \; \} \tag{1}$$

where the index i defines the operating condition in which the coefficient is used. C_l represents the lift coefficient, C_d is the drag coefficient, C_m is the momentum coefficient and Φ is a function of the pressure coefficient C_p.

Constraints on the airfoil shape and on its aerodynamic characteristics are used. The aerodynamic constraints are used to control the values of the design coefficients (typically C_l or C_m), while geometric constraints are used both to satisfy structural requirements and to indirectly control aerodynamic characteristics in off-design conditions. For instance, stall due to leading edge separation or loss of leading edge suction are related to the value of the ordinates at the airfoil leading edge. The following expression is used for the constraints:

$$G = K(\frac{V}{\hat{V}} - 1) \tag{2}$$

where V is an aerodynamic coefficient or a geometric parameter and $\hat{V}$ is a limit imposed on it. K represents a multiplication factor which is used to give a relative weight to each equation of the constraint set; in general, it is set to be equal to 1.

Minimization algorithm

Some optimization routines coming from the commercial package ADS [1] have been selected. For unconstrained problems, we usually use a quasi-Newton method like the Broydon-Fletcher-Goldfarb-Shanno (BGFS) variable metric method [2]; while for constrained problems we have selected a method based on the feasible direction algorithm [3]. We here describe in some details the feasible direction method which is perhaps not so familiar.

At each iteration q of the minimization process, the design vector $\mathbf{X}$ is updated according to the formula $\mathbf{X}^q = \mathbf{X}^{q-1} + \omega \mathbf{S}^q$, where $\mathbf{S}^q$ is a unit vector representing a search direction in a space having as many dimensions as the design variables, and where ω defines the displacement in the direction $\mathbf{S}^q$. The search direction is built in such a way that it will reduce the objective function without violating the constraint for some finite move, and thus can be defined as a "constrained steepest descent" direction.

Any vector $\mathbf{S}$ which reduces the objective function is called a usable direction. Clearly, such a direction will make an angle greater than 90° with the gradient vector of the objective function (see Fig.3). This suggests that the dot product of $\nabla F(\mathbf{X})$ and $\mathbf{S}$ should be negative. The limiting case is when the dot product is zero, in which case the vector **S** is tangent to the plane of constant objective function. Mathematically, the usability requirement becomes

$$\nabla F(\mathbf{X}) \cdot \mathbf{S} \leq 0 . \tag{3}$$

A direction is called feasible if, for some small move in that direction, the active constraint will not be violated. That is, the dot product of $\nabla G_j(\mathbf{X})$ with **S** must be nonpositive. Thus

$$\nabla G_j(\mathbf{X}) \cdot \mathbf{S} \leq 0 . \tag{4}$$

The search direction **S** is then determinated by considering the following problem:

Minimize:

$$\nabla F(\mathbf{X}) \cdot \mathbf{S} . \tag{5}$$

Subject to:

$$\nabla G_j(\mathbf{X}) \cdot \mathbf{S} \leq 0 \tag{6}$$

$$\mathbf{S} \cdot \mathbf{S} \leq 1 . \tag{7}$$

The last inequality has been introduced to ensure that the search vector remains bounded. Solving this problems gives a search direction which is tangent to the critical constraint boundaries, unless the objective can be reduced more rapidly by moving away from one or more constraints.

In the optimization code, the gradients are calculated by a first order forward finite difference unless a variable is at its upper bound. In this case, a first order backward finite difference step is used.

Flow solver

The optimization routine is coupled with an aerodynamic code which gives the values of the aerodynamic coefficients in order to calculate their derivatives with respect to the design variables. Based on these derivatives, the optimization routine chooses the most suitable modification direction. The more reliable the aerodynamic code is, the more credible the design. But, since a large number of aerodynamic coefficients evaluations is expected, it is also important to use a low-computational-cost aerodynamic method whose formulation is able to treat transonic flows. We have chosen a code that solves the full potential equation in the subsonic and transonic range by a finite difference technique and includes a boundary layer routine [4]. Thus, some viscous flows can be computed.

The physical domain around the airfoil is mapped onto the region inside a circle of unit radius and the physical infinity is placed at its center. The airfoil contour is transformed into the unit circle using Sells conformal mapping technique. The full potential equation

$$\left(1 - \frac{\phi_x^2}{a^2}\right)\phi_{xx} - 2\frac{\phi_x \phi_y}{a^2}\phi_{xy} + \left(1 - \frac{\phi_y^2}{a^2}\right)\phi_{yy} = 0 \ , \tag{8}$$

where a is the sound speed, is then solved using a quasi-conservative method and a line overrelaxation technique. We estimate that the computation of a transonic flow past an airfoil with a grid of about 5000 nodes requires approximately 10 seconds of CPU time on a CDC 910/600 workstation.

A boundary layer routine based on the integral method of Nash and MacDonald is used to simulate the viscous effects. The boundary layer is coupled to the inviscid solver in an iterative fashion. The coupling procedure adds the boundary layer thickness to the airfoil surface so that a new airfoil is obtained and the inviscid flow can be computed again. No special treatment is provided neither for shock/boundary layer interaction nor for separated regions, so only attached or weakly separated flows are computed.

Shape definition of the airfoil

The airfoil geometry is built using an approach similar to the one introduced in [5]. The shape is defined by adopting splines whose control parameters are represented by a set of design variables $\mathbf{X}$. They are split into a global and a local part, i.e. $\mathbf{X} = (\mathbf{X}^{glob}, \mathbf{X}^{loc})$.

The global part specifies the location of the airfoil with regard to a global coordinates system. It consists in four parameters $\mathbf{X}^{glob} = (l_x, l_y, \theta, c)$. (l_x, l_y) locates the leading edge (l_x is kept fixed and equal to zero), θ sets the chord angle (rotation about the leading edge) and c scales the chord. The local part defines the shape of the airfoil. This is done in terms of a spline approximation. Let $\mathbf{r}(t)$ be the position vector of the curve in the local frame, then we set

$$\mathbf{r}(t) = \sum_{i=0}^{n} B_{i,n}(t)\bar{\mathbf{r}}_i \quad ; \quad 0 \leq t \leq 1 \tag{9}$$

where $\bar{\mathbf{r}}_i$ is the position vector of the ith vertex of the defining curve polygon and $B_{i,n}(t)$ is a general basis function. Two kinds of basis functions have been investigated. The first is based on the so-called Bézier curve segment approximation for which $B_{i,n}(t)$ represent the Bernstein basis functions:

$$B_{i,n}(t) = \binom{n}{i} t^i (1-t)^{n-i} \tag{10}$$

In the second case, $B_{i,n}(t)$ are cubic B-spline functions [6]. The above expansion is done separately for the lower and upper sides of the airfoil. The number n and the abscissae $\bar{x}_i$ of the control polygon vertices are given in input. The deviations of the vertices ordinates from the initial ones $(\bar{z}_i)$ represent design variables:

$$\bar{\mathbf{r}}_i = (\bar{x}_i, \bar{z}_i + X_i \bar{z}_i) . \tag{11}$$

As the x-locations of the control vertices are specified, we have first to invert the equation $x = x(t)$ in order to calculate the corresponding t-parameter values. This has to be done only once, because the x-locations are kept fixed during calculations. The ordinates $\bar{z}_i$ can then be obtained by imposing that the curve passes through $n + 1$ points of the initial shape.

Some geometric properties of the above spline representations follows:

- The curve passes through the first and the last knots of the control polygon which correspond to the parameter values $t = 0$ and $t = 1$, respectively.
- The tangent vectors at both ends of the curve are parallel to the straight lines joining the function value at the end point and at the neighbouring interior point.
- The number of intersection between an arbitrary straight line and a Bézier segment curve or a B-spline curve is not greater than the number of intersections between that straight line and the curve defining polygon. This the variation diminishing property.

- For a B-spline curve, if a certain control polygon vertex is varied, only the part of the curve in the immediate neigbourhood of that vertex is affected. A Bézier segment curve does not have this property; if one part of it is changed, the effect extends over the whole curve.

Fig. 4 illustrates the representation of a generic airfoil using the two approaches, and of the corresponding defining curve polygons. In the optimization procedure, we prefer to use the Bézier curve segment approximation because changes in the position of a control polygon vertex often maintain a smooth curve, which is not always the case with B-splines.

NUMERICAL EXPERIMENTS

In this section, we present some results for the computation of test case T6 which has been defined within the workshop associated to BRITE/EURAM AERO-0026 contract. We consider the following problem :

$$\text{Minimize } C_D$$
$$\text{under the constraint of a given lift } C_l^t.$$

The freestream Mach number is taken to be $M = 0.73$ and the angle of attack is $\alpha = 2°$. The value of the target lift coefficient is $C_l^t = 1.05$. The initial profile is the RAE2822 airfoil, whose C_p distribution at the above conditions is shown in Fig. 5. Table 1 summarizes its performances. Two different ways to perform this problem has been investigated.

Table 1: Performances of the RAE2822 airfoil.

	C_L	C_D	thickness
Full potential sol.	0.875	0.0032	0.120

We first used a penalty method; that means that we minimize an augmented objective function without constraints:

$$C_D + \beta(C_l - C_l^t)^2 \tag{12}$$

where β is a parameter. The minimization is performed using the BFGS variable metric method. Figs. 6 and 7 illustrate the solutions obtained with $\beta = 1$ and $\beta = 100$, respectively. The figures show the C_p distribution of the new airfoils, together with the convergence history of the objective function as a function of the number of elementary problems. Table 2 summarizes the performances of the "optimized" airfoils.

Table 2: Performances of the "optimized" airfoils using the penalty method.

Value of β	C_L	C_D	thickness
1	1.046	0.0026	0.076
100	1.047	0.0076	0.099

We see that the solution depends on the value of the parameter β. This emphasizes two general problems which are present in optimization: the first, is the way to build the objective function and, the second, to understand whether an objective function has an absolute optimum or, on the contrary, if relative or very flat minima are present. As a matter of fact, with $\beta = 1$ the first term in Eq. (12) rapidly prevails when the value of lift draws near to the target; while for $\beta = 100$, the relative weight of the drag is reduced. The two objective functions should lead to the same optimum, but relative minima possibly exists, and are certainly different for the two functionals. In the optimization of airfoils, it is quite difficult to understand a priori how the objective function looks like or to evaluate the effect of the initial condition on the solution. A solution computed with an Euler solver on the optimized shape corresponding to $\beta = 1$, indicates that the two solutions are identical in the windward part of the airfoil while they show a different supersonic recompression after the suction peak, and that the full potential solution underestimates the shock intensity and consequently the value of the wave drag. The Euler solution is illustrated in Fig. 8. The performances computed using the two solvers are summarized in Table 3.

Table 3: Comparison of the performances of the "optimized" airfoil using two different solvers.

	C_L	C_D	thickness
Full potential sol.	1.046	0.0026	0.076
Euler sol.	1.039	0.0052	0.076

Secondly, we used the capability of the optimizer to treat constrained problems in order to minimize the original objective function with the nonlinear equality constraint:

$$1 - \frac{C_l}{C_l^t} = 0. \tag{13}$$

The basic concept is to first change the sign on the constraint, if necessary, so that the scalar product of the gradient of the constraint with the gradient of the objective function is negative. The constraint is then converted to a non-positive inequality constraint and a linear penalty function is added to the objective function. The penalty function, together with the conversion to an inequality constraint have the effect of driving the original equality constraint to zero at the optimum. With this approach, the numerical procedure minimizes drag and tries to satisfy "at the best" the equality constraint. We are a priori not sure that the target lift will be satisfied, but we can expect a relative error of a few per thousands. The solution obtained using this approach is illustrated in Fig. 9, together with the histories of lift and drag coefficients. In comparison with the previous solutions, the present one reduces the wave drag increasing the suction peak. Note that the supersonic recompression after the suction peak remains very smooth. As previously showed, the Euler solution, which is illustrated in Fig. 10, indicates that the full potential solution we have computed underestimates the intensity of the shock and consequently the value of the wave drag. But, it is important to note that the shape which corresponds to the smallest value of the drag computed by the full potential solver is also that which leads to the smallest value obtained with an Euler solver. Consequently, the full potential solver gives the right trend and can be profitably used in an optimization loop. The performances of the new airfoil computed using the two solvers are summarized in Table 4.

Table 4: Comparison of the performances of the "optimized" airfoil using two different solvers.

	C_L	C_D	thickness
Full potential sol.	1.047	0.0016	0.087
Euler sol.	1.049	0.0029	0.087

The heaviest airfoil optimization we have performed costs about 40 minutes of CPU time on a CDC 910/600 workstation.

CONCLUSIONS

The airfoil optimization procedure we have developed is able to treat optimization problems such as to minimize drag, maximize lift, recover a target pressure distribution, etc. . An important feature of the method is that non linear constraints can be taken into account in order to satisfy structural requirements and to indirectly control aerodynamic characteristics in off-design conditions. The optimization code is cheap (less than 1 hour of CPU time on a CDC 910/600 workstation to perform a complete optimization loop which requires dozens of aerodynamic computations). Even if the solver underestimates the inviscid wave drag (by comparison with an Euler solver), it nevertheless gives the right trend and can be profitably used in an optimization procedure.

AKNOWLEDGMENTS

This work has been carried out under the BRITE/EURAM contract AERO-0026 funded by CEC-DGXII Research Directorate.

REFERENCES

[1] G.N. Vanderplaats: ADS - A Fortran Program for Automated Design Synthesis, User Manual, 1987.

[2] C.G. Broydon: The Convergence of a Class of Double Rank Minimization Algorithms, Part I and II, J. Inst. Maths. Applns., Vol 6, 1970, pp. 76-90 and 222-231.

[3] M. Zoutendijk: Methods of Feasible Directions, Elsevier Publishing Co, Amsterdam, 1960.

[4] F. Bauer, P. Garabedian, D. Korn and A. Jameson: Supercritical Wing Sections, Lecture Notes in Economics and Mathematical Systems, Springer-Verlag, 1972.

[5] H. Schwarten: Wing Design with 3D-Subsonic Inverse Panel Method, BRITE-EURAM Contract AERO-0026 Mid-Term Report, 1991.

[6] G. Farin: Curves and Surfaces for Computer Aided Geometric Design, Academic Press, 1989.

DESIGN
SPECIFICATIONS
STARTING
AIRFOIL
AIRFOIL DESIGN
PROCEDURE
FINAL
AIRFOIL
MODIFIED
AIRFOIL 1
. . . .
MODIFIED
AIRFOIL N

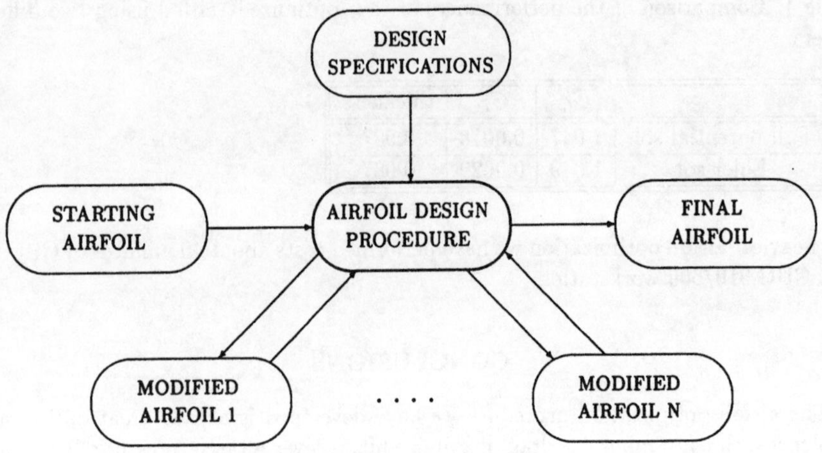

Fig.1: Iterative design process.

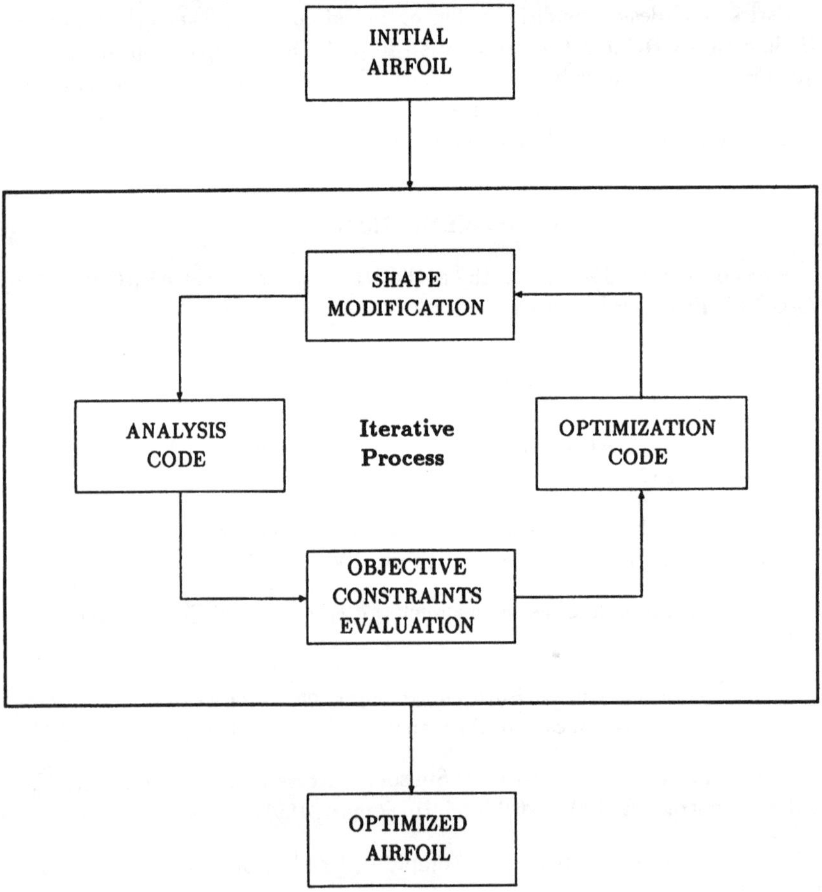

INITIAL
AIRFOIL
SHAPE
MODIFICATION
ANALYSIS
CODE
Iterative
Process
OPTIMIZATION
CODE
OBJECTIVE
CONSTRAINTS
EVALUATION
OPTIMIZED
AIRFOIL

Fig.2: Aerodynamic design via optimization.

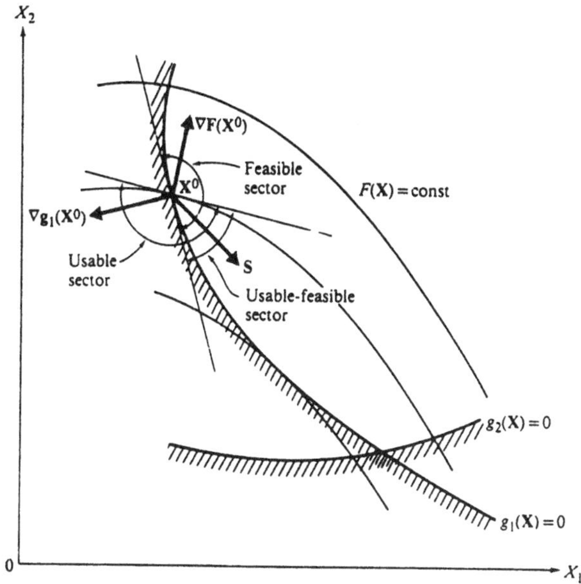

Fig.3: Usable-feasible search direction.

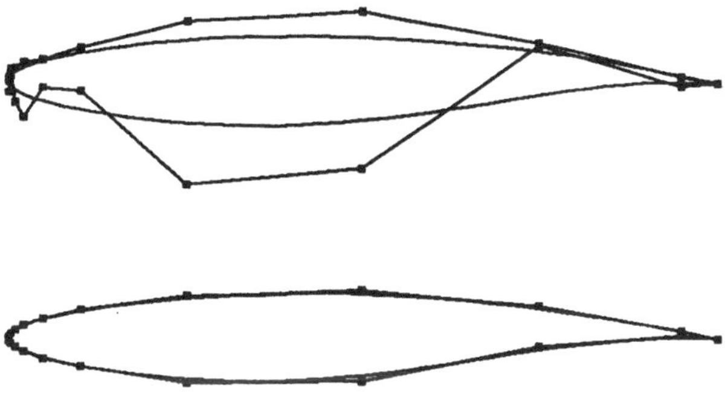

Fig.4: Shape approximation using Bézier segments (upper) and B-splines (lower).

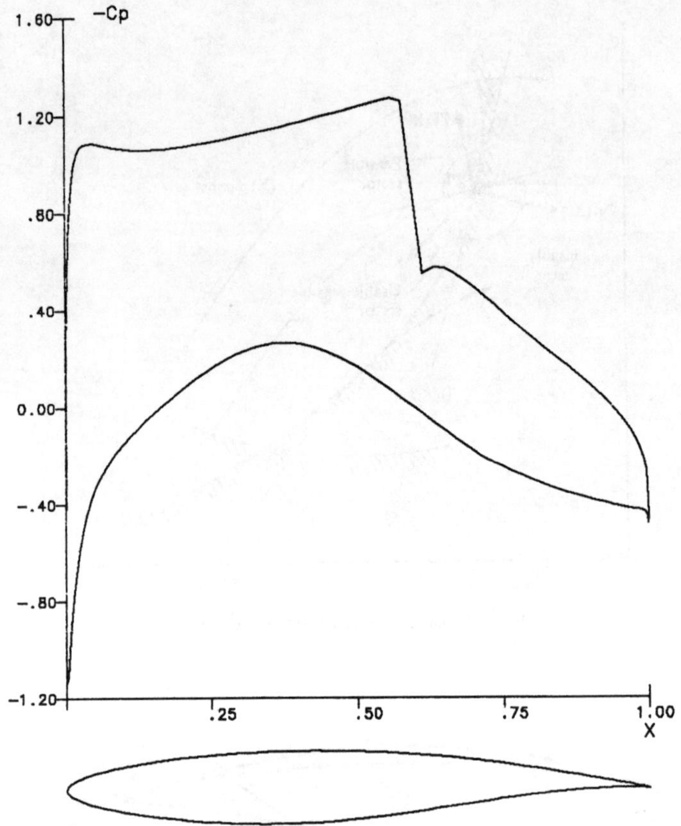

Fig. 5: C_p distribution for RAE2822 airfoil.

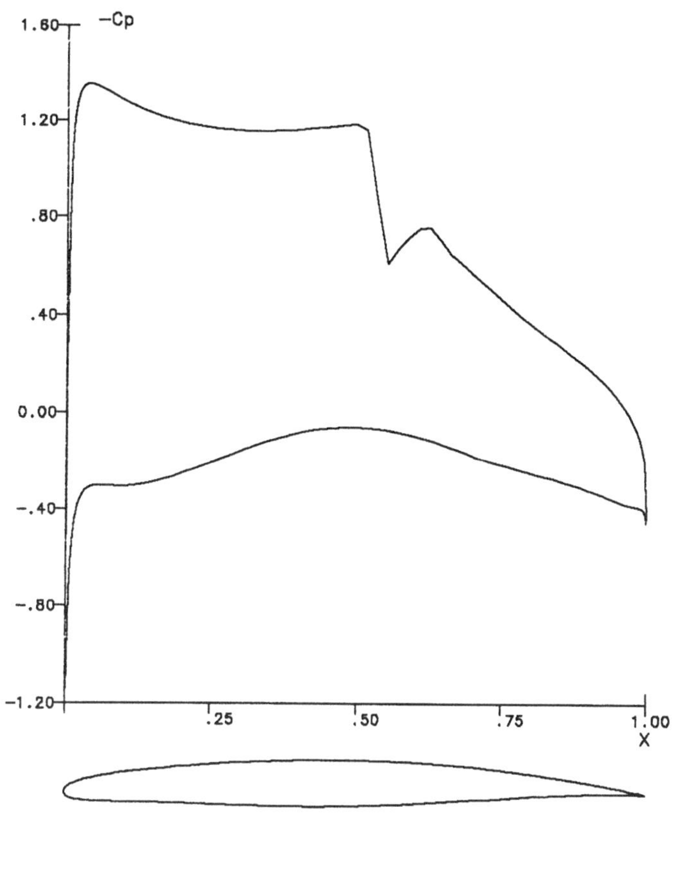

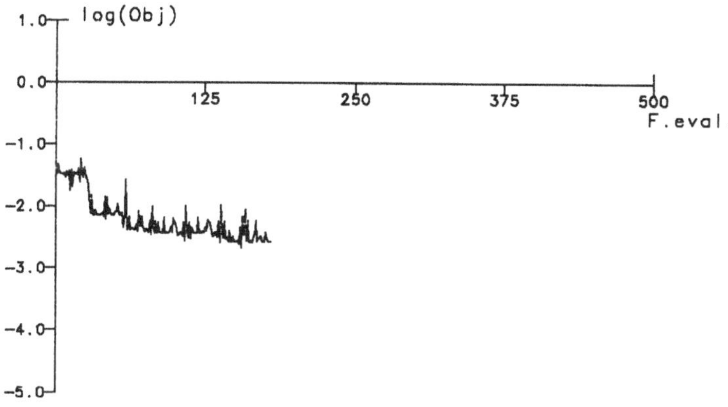

Fig. 6: Optimized airfoil using a penalty function ($\beta = 1$).
C_p distribution and convergence history.

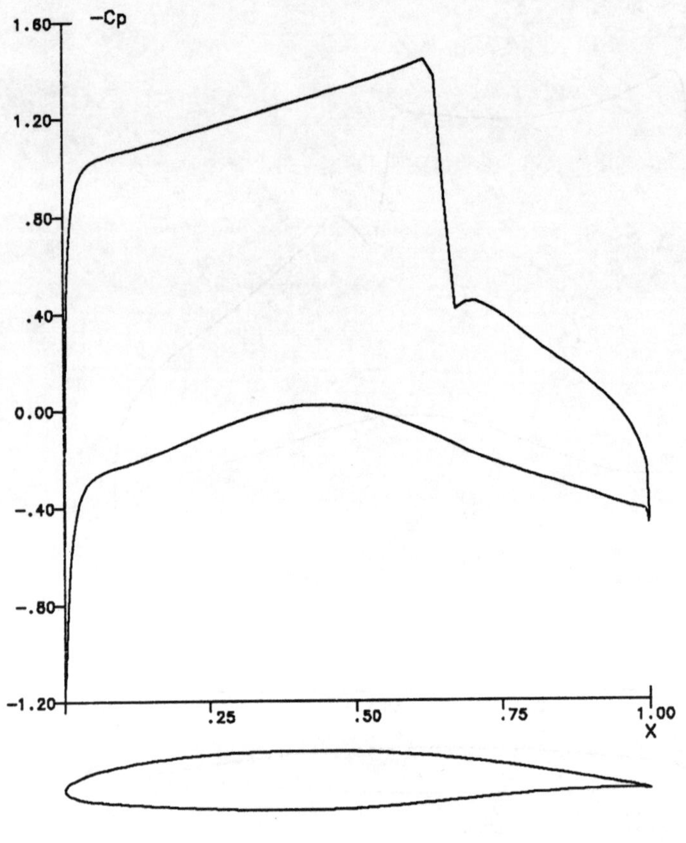

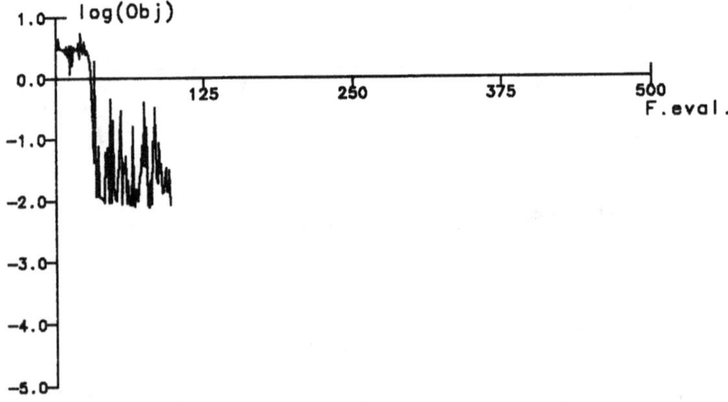

Fig. 7: Optimized airfoil using a penalty function ($\beta = 100$). C_p distribution and convergence history.

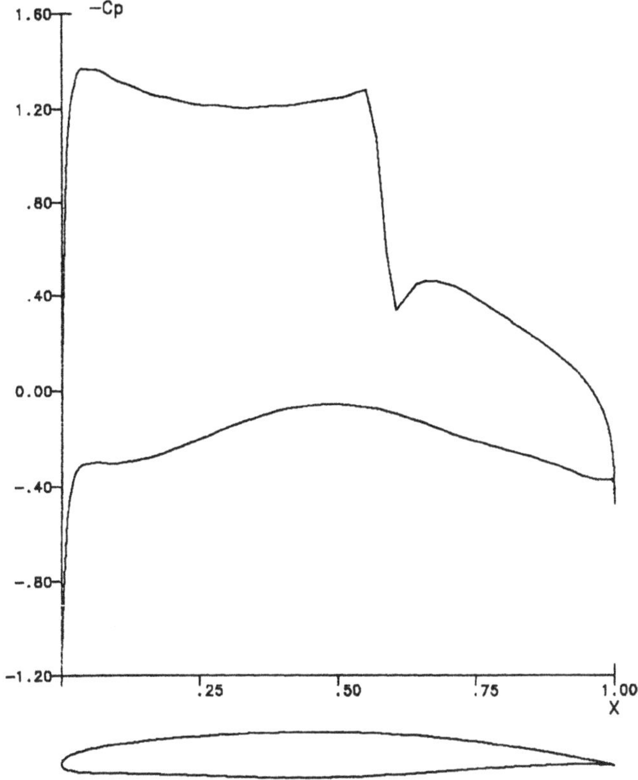

Fig. 8: C_p distribution computed using an Euler solver. The shape is the one of Fig. 6.

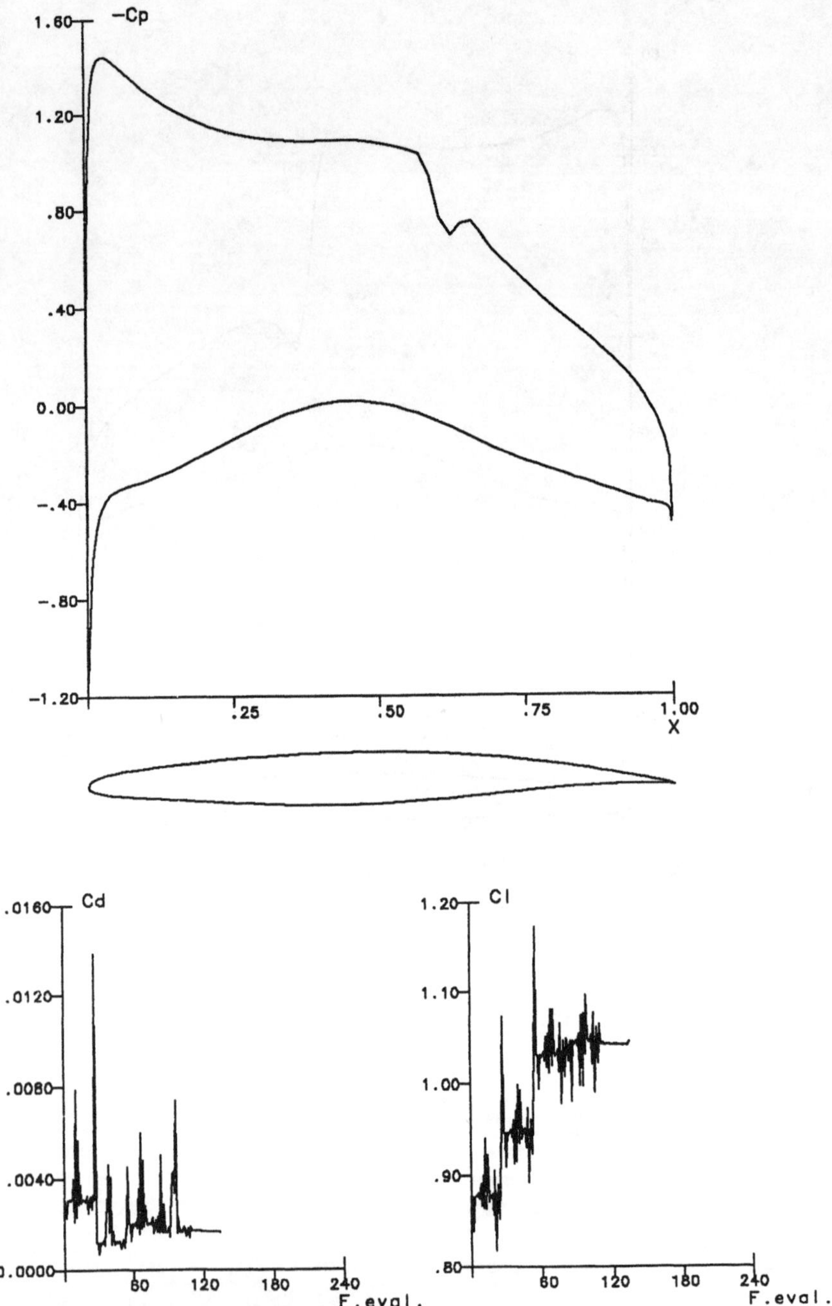

Fig. 9: Optimized airfoil using the approach based on the equality constraint. C_p distribution and aerodynamic coefficients histories.

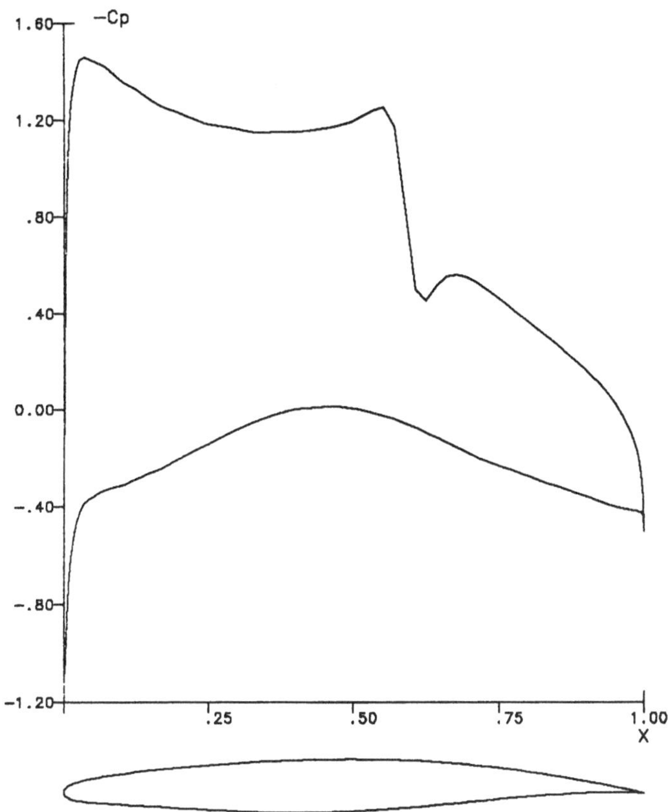

Fig. 10-a: C_p distribution computed using an Euler solver. The shape is the one of Fig. 9.

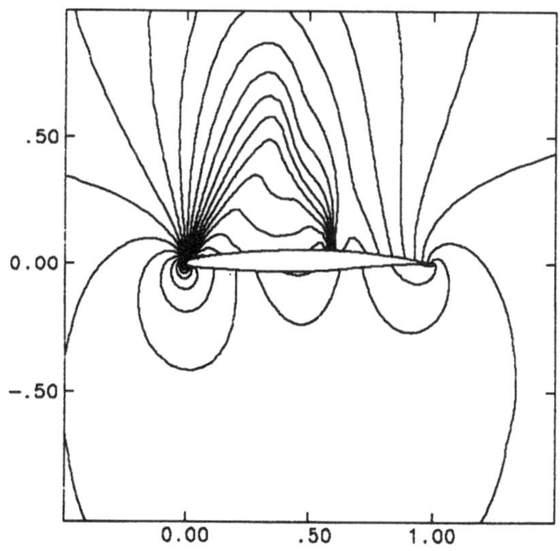

Fig. 10-b: Iso-C_p contours for the Euler computation.

A RESIDUAL CORRECTION METHOD APPLIED TO 2D MULTI-POINT AIRFOIL DESIGN AND 3D SINGLE-POINT WING DESIGN

Th.E. Labrujère and J. van der Vooren
National Aerospace Laboratory NLR
P.O. Box 90502, 1006 BM Amsterdam, The Netherlands

SUMMARY

Results are presented of a residual correction method to solve the multi-point airfoil design problem for subsonic and transonic conditions. The method is based on full-potential theory and minimizes a cost function weighting the deviations from specified target pressure distributions for each design condition in a least squares sense. Deviations from specified target pressure distributions (residuals) are translated into airfoil geometry corrections, by means of an approximate inverse calculation using a linearized panel method. The feasibility of introducing an equivalent incompressible multi-point design problem is demonstrated for a two-point reconstruction test case. First results of a two-point example airfoil design case are presented. The applicability of the residual correction methodology to 3D single-point wing design is demonstrated as well.

INTRODUCTION

Nowadays, computational methods are widely accepted by aerodynamic designers as tools to analyze aerodynamic characteristics of given aircraft components or even complete aircraft. In the development of these methods considerable progress has been made with respect to complex flow analysis capabilities. As a consequence, comparison of alternatives during the process of designing aerodynamic shapes is increasingly supported by applying analysis methods. The design process itself, however, i.e. the definition of possible alternatives, may be improved further by the development of computational design methods for the determination of aerodynamic shapes, aiming at specific aerodynamic targets.

Computational design methods have not yet reached a stage of development comparable with that of the analysis methods. Conformal mapping was used in the earliest design methods for the solution of the 2D inverse problem for the design of an airfoil in incompressible flow, generating an airfoil shape that produces a prescribed pressure distribution (Lighthill [1];Timman [2]). Since then, considerable progress has been made in solving similar problems for more complex flows. For instance, Drela [3] solved the 2D inverse problem for the Euler equations by applying an iterative Newton method .

However, optimum design of 2D airfoils satisfying requirements for more than one operating condition and taking into account essential geometric constraints is only in the early stages of development. And, e.g. automatic procedures for solving the same problem in 3D wing design are even further remote.

Many methods for solving the aerodynamic design problem are based on the residual correction approach. The residual, formulated in terms of the difference between current and target pressure distributions, is used to define new estimates of the aerodynamic shape by applying an approximate inverse calculation procedure. The current pressure distribution for a given estimate of the aerodynamic shape is determined by means of a flow analysis method.

Within the BRITE/EURAM project "Optimum design in aerodynamics" an algorithm for computational multi-point wing design has been developed at NLR [4]. Here, results are presented of a 2D method based on this algorithm and applied to the problem of multi-point optimum design of airfoils in transonic flow, using full potential theory. The method comprises basically two major computational loops.

In the outer loop, for each operating condition considered, the differences between the current and target pressures in transonic flow are translated into a target pressure difference in an equivalent incompressible flow.

In the inner loop, considering the realization of all such target pressure differences as an equivalent incompressible multi-point design problem, a new estimate of the airfoil geometry is determined by applying a least squares technique for the solution of a linearized flow problem.

The approach is in favour of computational speed, because the translation to an equivalent incompressible problem allows fast integral equation methodology (panel methods) to be used.

The choice for the residual correction approach applied to an equivalent incompressible problem is motivated in view of good experiences at NLR in transonic single-point airfoil and wing design (Fray et al [5]; Brandsma et al [6]). The feasibility of this approach for the solution of the 2D airfoil design problem in transonic flow is demonstrated for a single-point as well as a two-point reconstruction test on a Korn airfoil.

As an example of multi-point airfoil design a two-point viscous subsonic/transonic airfoil design problem is considered. This example concerns the design of an airfoil which combines favourable high speed and low speed performance. For each of the two operating conditions, defined by angle of attack and Mach number, target pressure distributions are specified. Finally, an example of 3D wing design is used to demonstrate the applicability of the residual correction approach for the 3D wing design problem in subsonic flow.

MULTI-POINT AIRFOIL DESIGN

The multi-point airfoil design problem is defined as the minimization of the cost function

$$F = \frac{1}{2}\sum_{i=1}^{n} W_i \int_{chord} [C_p^i(x) - C_{p_{tar}}^i(x)]^2 dx$$

where the summation is over the n operating conditions, W_i is a weight factor that balances the operating conditions , C_p^i is the $C_{p_{tar}}^i$ is the target pressure distribution.

The aerodynamic designer must specify each target pressure distribution as the balanced result of desired aerodynamic characteristics and of aerodynamic constraints that are better not violated. It has been concluded at NLR that specification of the cost function in this way is the most feasible.
In addition, the aerodynamic designer generally needs a possibility to specify certain geometric constraints, originating either from aerodynamic or structural arguments. Both equality and inequality constraints can occur. Examples are bounds on trailing edge angle and thickness, and a specified thickness to chord ratio. It is assumed that the airfoil contour will be a closed curve of which the coordinates will be given by means of x and z (see Fig.1).

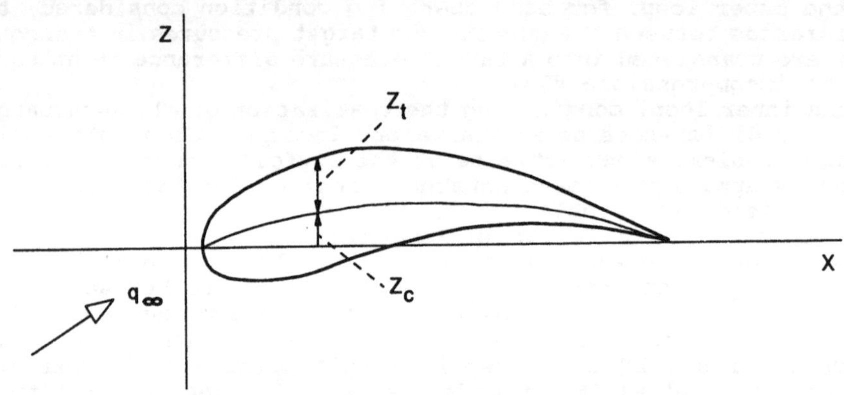

Fig. 1 Airfoil axis system

RESIDUAL CORRECTION METHOD

The residual correction method uses basically the following computational outer loop in an iterative fashion to define new estimates for the airfoil geometry :
1. Calculate the full-potential flow about the current estimate of the airfoil geometry z=z[x] for each operating condition (or design point), using an (available) transonic analysis code, and obtain the current pressure distribution C_p[x] on the airfoil surface.
2. Decide whether or not the current airfoil geometry z=z[x] needs further improvement, by comparing the current pressure distribution C_p[x] with the target pressure distribution $C_{p_{tar}}$[x] for each operating condition, and by considering the convergence history.
3. If further improvement of the current airfoil geometry z=z[x] is considered necessary, calculate the equivalent incompressible perturbation velocity defect δu on the airfoil contour, for each operating condition, from the target pressure distributions $C_{p_{tar}}$ [x] and the current pressure distributions C_p[x].

4. Calculate the airfoil geometry correction $\delta z=\delta z[x]$ using the linearized inverse panel method described in the relevant section below, obtain a new estimate for the airfoil geometry from

$$z[x] \Rightarrow z[x] + \delta z[x], \tag{2}$$

and go back to step 1. The flow diagram of this computational outer loop is given in Fig. 2. Details of step 3 are discussed below.

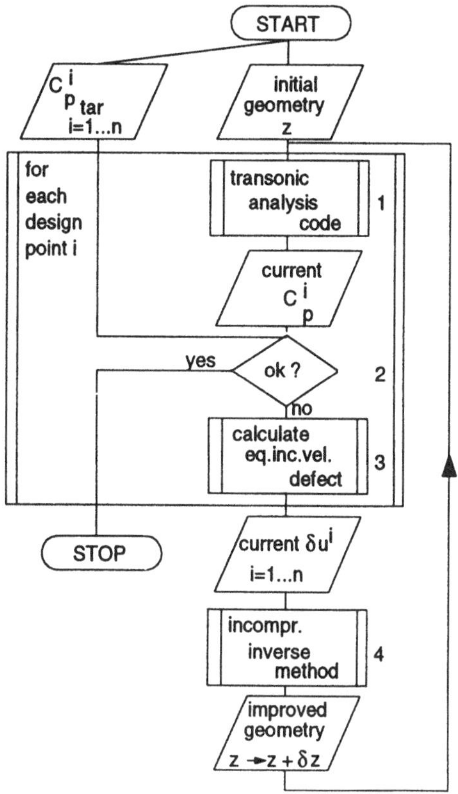

Fig. 2 Residual correction method, outer loop

Pressure defect splitting

A splitting technique is used to distinguish between the subsonic and the supersonic part of the pressure defect. The split made is based primarily on the assumption that subsonic thin-airfoil theory is applicable if the local actual pressure and the local target pressure are both subsonic, and that supersonic wavy-wall theory is applicable if both pressure coefficients are supersonic. In case one of them is of a different nature, the critical pressure coefficient is used as upper or lower limit. In [5] a detailed description of the derivations is given; here only the resulting formulas are presented.

The subsonic and supersonic parts of the pressure defect δC_p are defined as

$$\delta_{sup} C_p = \min(C_{p_{tar}}, C_p^*) - \min(C_p, C_p^*) \,,$$

$$\delta_{sub} C_p = \max(C_{p_{tar}}, C_p^*) - \max(C_p, C_p^*) \,, \tag{3}$$

where

C_p is the actual pressure coefficient,

$C_{p_{tar}}$ is the target pressure coefficient and

C_p^* is the critical pressure coefficient.

The procedure is illustrated in Fig.3.

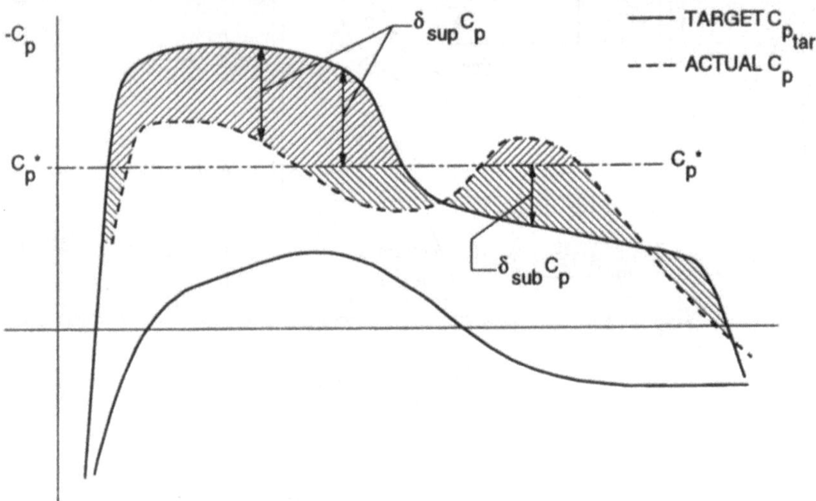

Fig. 3 Pressure defect splitting

Equivalent incompressible perturbation velocity defect pertaining to thin-airfoil theory.

From the <u>subsonic</u> part of the pressure-defect distribution an equivalent incompressible perturbation velocity defect pertaining to thin-airfoil theory is derived by means of the following relation,

$$\delta_{sub} u = \min(u_{tar}, u^*) - \min(u, u^*) \,. \tag{4}$$

Here, u_{tar}, u^* and u are derived from the general relations Note the inverse Riegels-type leading edge correction in the expression for u.

$$u = \beta[q*\sqrt{1+(\frac{1}{\beta}\frac{dz}{dx})^2} - 1] ,$$

$$\beta = \sqrt{1 - M_\infty^2} , \qquad (5)$$

$$q^2 = 1 - [\{1 + \frac{\gamma}{2} M_\infty^2 C_p\}^{\frac{(\gamma-1)}{\gamma}} - 1] / [\frac{(\gamma-1)}{2} M_\infty^2] .$$

From the <u>supersonic</u> part of the pressure defect distribution an incompressible perturbation velocity defect is derived by means of the set of equations

$$\delta_{sup} u_u = \delta_e u_t + \delta_e u_c \quad \text{on the upper (u) side of the airfoil ,}$$

$$\delta_{sup} u_l = \delta_e u_t - \delta_e u_c \quad \text{on the lower (l) side of the airfoil ,} \qquad (6)$$

with

$$\delta_e u_t(x) = \frac{1}{\pi} \int_{x_{le}}^{x_{te}} \{\delta_{sup} \frac{dz_t}{d\xi}\} \frac{d\xi}{x - \xi} ,$$

$$\delta_e u_c(x) = \frac{1}{\pi} \sqrt{\frac{x_{te} - x}{x - x_{le}}} \int_{x_{le}}^{x_{te}} \{\delta_{sup} \frac{dz_c}{d\xi}\} \sqrt{\frac{\xi - x_{le}}{x_{te} - \xi}} \frac{d\xi}{x - \xi} , \qquad (7)$$

being equivalent incompressible velocity defects corresponding to the effects of airfoil "thickness" (t) and "camber" (c).

Here

$$\delta_{sup} \frac{dz_t}{dx} = \frac{1}{2} (\delta_{sup} \frac{dz_u}{dx} - \delta_{sup} \frac{dz_l}{dx}) ,$$

$$\delta_{sup} \frac{dz_c}{dx} = \frac{1}{2} (\delta_{sup} \frac{dz_u}{dx} + \delta_{sup} \frac{dz_l}{dx}) , \qquad (8)$$

with

$$\delta_{sup} \frac{dz}{dx} = \frac{1}{2} \sqrt{M_{loc}^2 - 1} \; \delta_{sup} C_p , \qquad (9)$$

$$M_{loc}^2 = \frac{2}{\gamma - 1} \left[\{1 + \frac{\gamma - 1}{2} M_\infty^2\} / \{1 + \frac{\gamma}{2} M_\infty^2 C_{p_{loc}}\}^{\frac{\gamma-1}{\gamma}} - 1 \right] , \qquad (10)$$

and

$$C_{p_{loc}} = \frac{1}{2} \left[\frac{1}{2} \{\min(C_p^t, C_p^*) + \min(C_p, C_p^*)\} + C_p^* \right] , \qquad (11)$$

being applicable for the upper(u) as well as the lower(l) side of the airfoil.

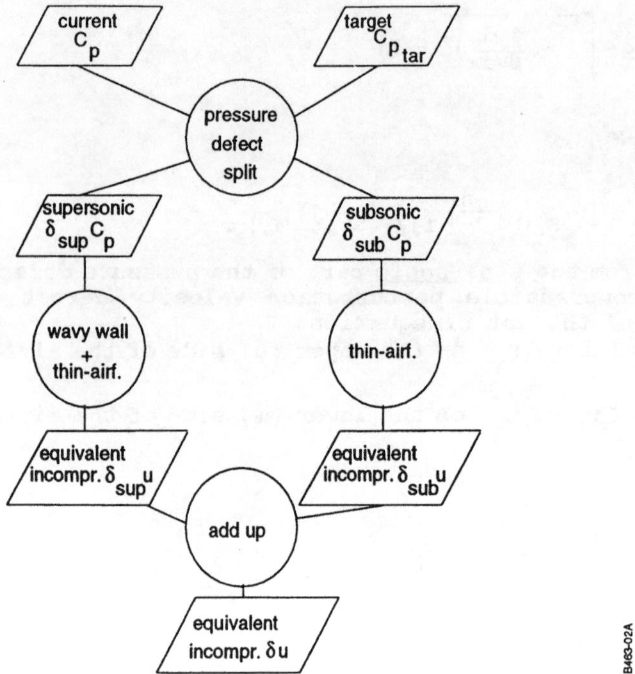

Fig. 4 Calculation of incompressible perturbation velocity defect

Finally the equivalent incompressible perturbation velocity defect is obtained by adding the subsonic and supersonic parts,

$$\delta u = \epsilon (\delta_{sub} u + \delta_{sup} u) \ , \tag{12}$$

where ϵ is a relaxation parameter.

An overview of the above described procedure to calculate the equivalent velocity defect is given in Fig.4.

Geometric corrections

Utilizing thin-airfoil theory and consequently splitting the velocity defect in a symmetrical and an anti-symmetrical part, the determination of the geometric corrections is formulated as a least squares problem. The solution to this problem is obtained by solving a system of linear algebraic equations for camber and thickness corrections.

Using the equivalent incompressible velocity defect distribution, symmetrical and anti-symmetrical perturbation velocities are defined for each design condition by

$$\delta u_t = \{\frac{1}{2} (\delta_{sub} u_u + \delta_{sub} u_l) + \delta_e u_t\} \epsilon ,$$

$$(13)$$

$$\delta u_c = \{\frac{1}{2} (\delta_{sub} u_u - \delta_{sub} u_l) + \delta_e u_t\} \epsilon ,$$

where ϵ is a relaxation factor.

The geometric corrections are related to the perturbation velocity distributions by the thin airfoil integral expressions

$$\delta u_t(x) = \frac{1}{\pi}\int_{x_{le}}^{x_{te}} \delta z_t(\xi)\,\frac{d\xi}{(x-\xi)^2}\,,$$

$$\frac{d\delta z_c}{dx}(x) = \frac{1}{\pi}\int_{x_{le}}^{x_{te}} \delta u_c(\xi)\,\frac{d\xi}{(\xi-x)}\,. \tag{14}$$

Constraints on the geometry have been limited to the requirement that the airfoil thickness and/or the camber will approximately satisfy prescribed values. Accordingly, the camber and thickness corrections are determined such that the following functional is minimized :

$$\begin{aligned}
\sum_i \Big\langle & \int_{x_{le}}^{x_{te}} w_u^i(x)\,\Big[\delta u_t^i(x) + \delta u_c^i(x) - \frac{1}{\pi}\Big\{\int_{x_{le}}^{x_{te}} \delta z_t(\xi)\,\frac{d\xi}{(x-\xi)^2} + \\
& \qquad + \sqrt{\frac{x_{te}-x}{x-x_{le}}}\int_{x_{le}}^{x_{te}} \frac{d\delta z_c}{d\xi}\sqrt{\frac{\xi-\xi_{le}}{\xi_{te}-\xi}}\,\frac{d\xi}{x-\xi}\Big\}\Big]^2 dx \\
& + \int_{x_{le}}^{x_{te}} w_l^i(x)\,\Big[\delta u_t^i(x) - \delta u_c^i(x) - \frac{1}{\pi}\Big\{\int_{x_{le}}^{x_{te}} \delta z_t(\xi)\,\frac{d\xi}{(x-\xi)^2} + \\
& \qquad - \sqrt{\frac{x_{te}-x}{x-x_{le}}}\int_{x_{le}}^{x_{te}} \frac{d\delta z_c}{d\xi}\sqrt{\frac{\xi-\xi_{le}}{\xi_{te}-\xi}}\,\frac{d\xi}{x-\xi}\Big\}\Big]^2 dx \, + \\
& + \sum_{k=1}^{K} w_{tk}\,[\delta z_t(x_k) - \delta z_t^t(x_k)]^2 + \\
& + \sum_{k=1}^{K} w_{ck}\,[\{\delta z_c(x_k) - \delta z_c(x_{i_k})\} - \{\delta z_c^t(x_k) - \delta z_c^t(x_{i_k})\}]^2 \Big\rangle\,.
\end{aligned} \tag{15}$$

The outer summation is over the different design conditions, indicated by the superscript i. The weight factors are given by w^i; the subscripts u,l,t,c refer to upper and lower side, to thickness and camber respectively.
The thickness is prescribed by means of its values in certain points; the camber is prescribed by means of the difference between the value in point k and the value in point i_k.
Utilizing a planar panel method with constant singularity strengths on the panels, the above functional is discretized. Putting the derivatives with respect to the unknown parameters of the expression thus obtained equal to zero, a linear system of algebraic equations is obtained. Its solution leads to corrections for thickness and camber.

COMPUTATIONAL RESULTS

Test case T4.

The translation of the transonic multi-point design problem into a sequence of equivalent incompressible multi-point design problems raises the question of feasibility.In order to demonstrate the feasibility, a single-point as well as a two-point reconstruction test case have been run. In both cases the NACA 64A410 airfoil has been specified as initial geometry.

For the single-point reconstruction test case both the inviscid as well as the viscous transonic pressure distribution were calculated for the Korn airfoil at zero angle of attack and Mach number 0.75. Hence, in fact two single-point test cases were defined by prescribing each of these pressure distributions as a target for the design process with the NACA airfoil as a start.

In the inviscid single-point case, application of the design method led to recovery of the Korn airfoil in about 25 iterations, each iteration involving one flow analysis calculation and one geometry update. For the viscous single-point case about 50 iterations were needed for recovery of the Korn airfoil.
In Fig.5 the pressure distributions on the reconstructed airfoil (inviscid and viscous) are compared with the target pressure distributions. The pressure distribution on the reconstructed airfoil deviates only slightly from the target, which reflects the almost complete recovery of the original airfoil. The convergence history of the process is presented in Fig. 5, where the value of the cost function and the value of the L_2 norm are given as a function of the number of iterations. Here the cost function is defined by :

$$F=\oint (C_{p_a}-C_{p_t})^2 ds \tag{16}$$

where the integration is along the airfoil contour, subscript "a" refers to the actual and subscript "t" to the target pressure distribution.

The L_2 norm is defined by :

$$L_2(z)=\int_{upper side}(z_a-z_t)^2 dx+\int_{lower side}(z_a-z_t)^2 dx\,. \tag{17}$$

In the multi-point case, inviscid pressure distributions calculated for a Korn airfoil at two different operating conditions (Figs.8,9) have been defined as target, and the NACA64A410 airfoil was specified as starting geometry. Application of the two-point implementation has indeed led to straightforward recovery of the target airfoil (Fig.6), as well as of the pressure distributions (Figs. 7,8).

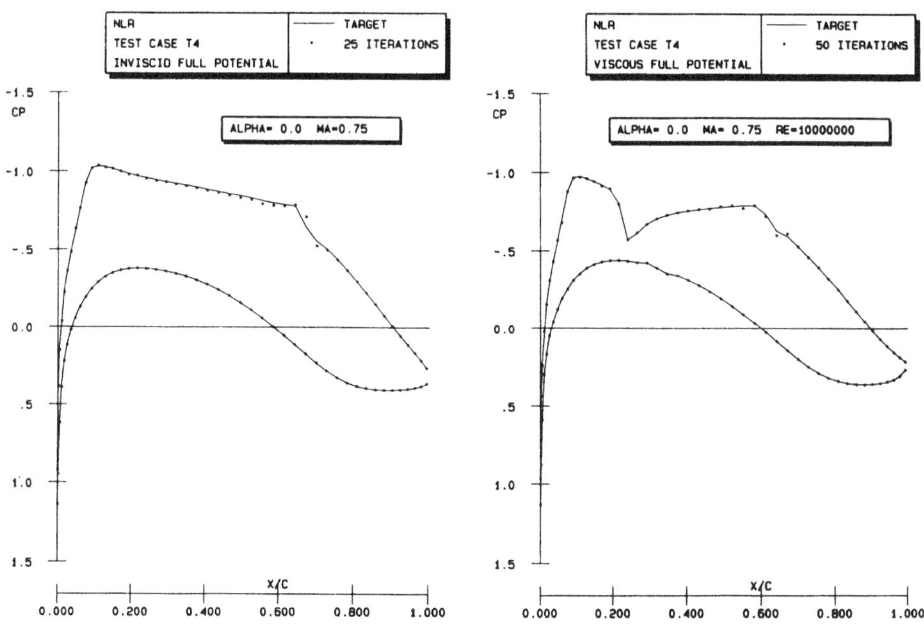

Fig. 5 Single-point reconstruction of Korn airfoil; pressure distributions

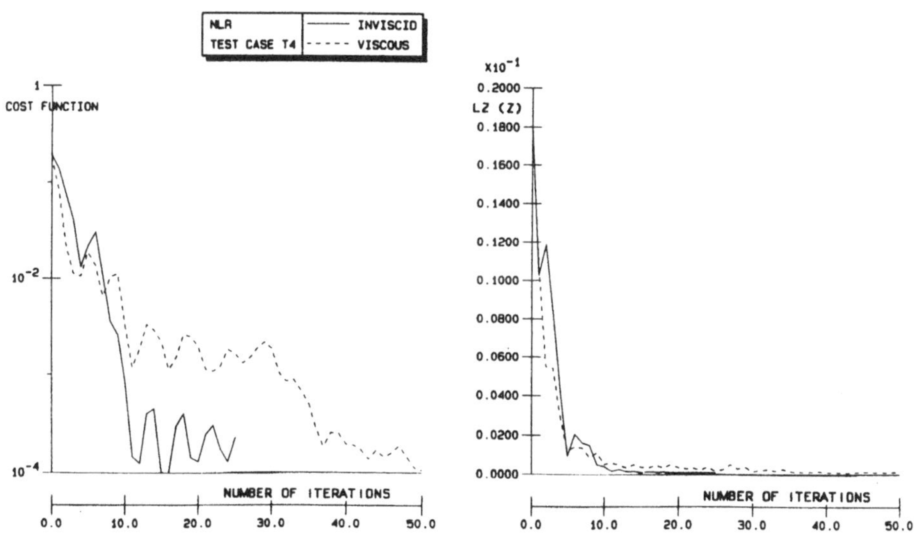

Fig. 6 Single-point reconstruction of Korn airfoil; convergence history

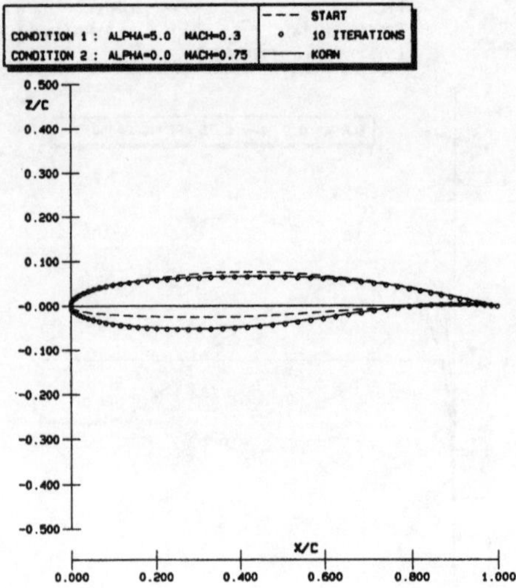

Fig. 7 Two-point reconstruction of Korn airfoil; geometry

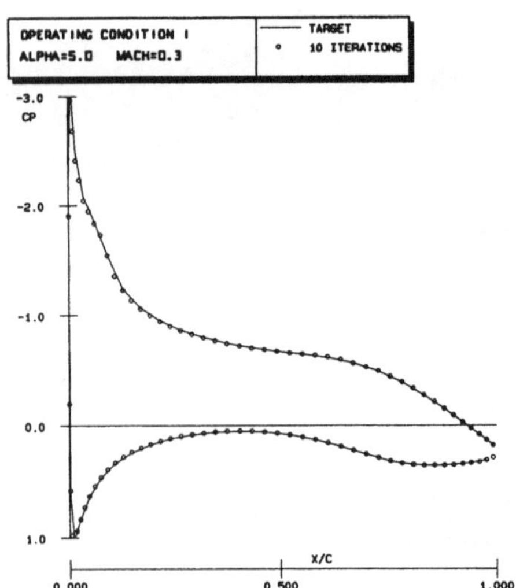

Fig. 8 Two-point reconstruction of Korn airfoil; pressure distribution

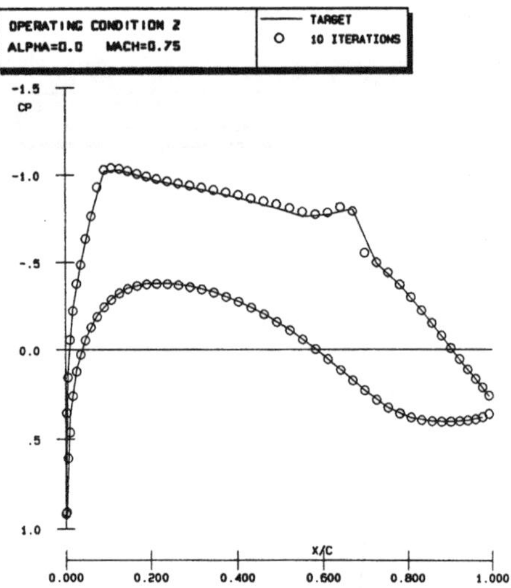

Fig. 9 Two-point reconstruction of Korn airfoil; pressure distribution

Test case T8.

An example two-point design case has been run to investigate the possibility to determine the geometry of an airfoil such that it combines favourable characteristics at two different operating conditions. Analogous design problems may be encountered e.g. at the design of a transport aircraft outer wing section (no flap, no slat) or at the design of a helicopter rotor blade section. The design problem was defined by prescribing two viscous target pressure distributions, one for Mach number 0.2 ($Re=0.5*10^7$) associated with a high lift capacity, and another for Mach number 0.77 ($Re=10^7$) chosen for its favourable high speed performance. Two weight factors were introduced to balance both requirements. Dealing with viscous target pressure distributions presents no special problem if a viscous full-potential interaction code is used in the computational outer loop of the residual correction method. Calculations have been performed for four choices of the weight factors showing the transition of a high lift airfoil (weight factor W_2 set equal to zero) into a high speed airfoil (weight factor W_1 set equal to zero). The NACA4412 airfoil has been used as starting geometry.

Fig.10 shows the different airfoil shapes obtained in this way and Fig.11 shows the corresponding pressure distributions. As such, the results clearly demonstrate that the design process in itself works. However, the present example can hardly be considered as a realistic multi-point design problem because of the choice of two completely incompatible design conditions. It may be expected that, in practice, these kind of problems will be of the re-design type, in which case the different design conditions are much more compatible and much more realistic results will be obtained.

The target pressure distributions prescribed in the present example have been obtained by means of a computational process in which the pressure distributions themselves were optimized with respect to desirable aerodynamic characteristics (one for high lift, one for low drag) without making a direct connection with the airfoil that should produce these pressure distributions. As a consequence, even in the single-point design cases (either W_1 or W_2 set equal to zero) there will remain a certain discrepancy between target and realized pressure distribution. Furthermore, the computational methods used for analysis and for geometry correction calculation have their own limitations. No attempts have been made to approximate the targets as close as possible. The results obtained were considered satisfactory with respect to the present demonstration.

Test case T14.

This test case is a single-point 3D wing design problem, which is used to demonstrate the applicability of the residual-correction approach. This design aims to improve the inviscid low speed high lift pressure distribution for a known wing-body configuration. Starting point is the DLR-F4 configuration for which a target pressure distribution is specified in a number of chordwise wing sections. The target has been obtained

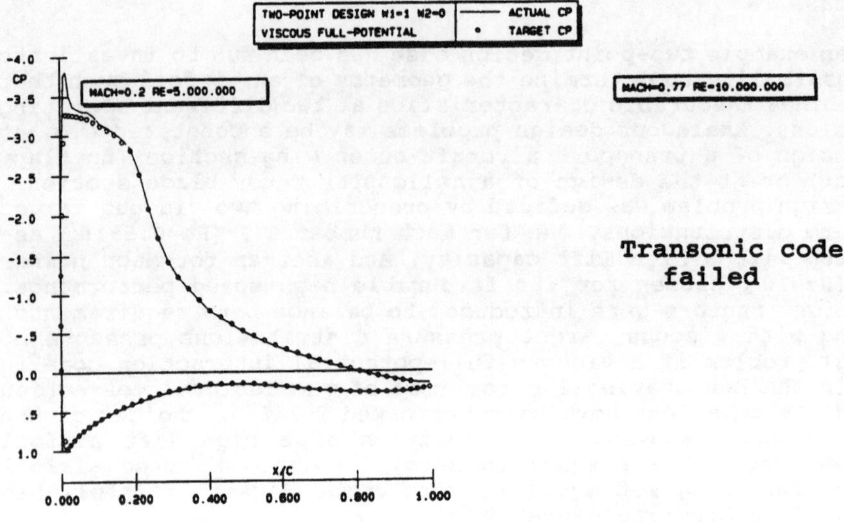

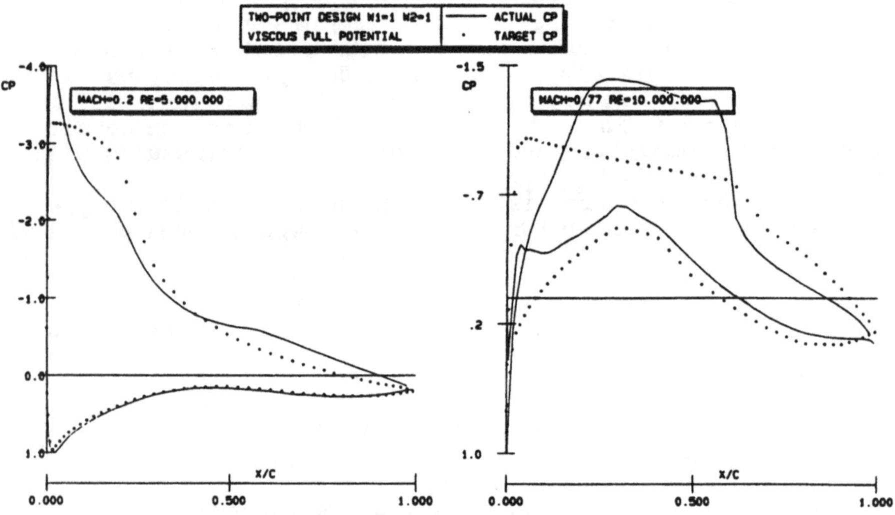

Fig. 11 Two-point airfoil design with different weight factor combinations; pressure distributions

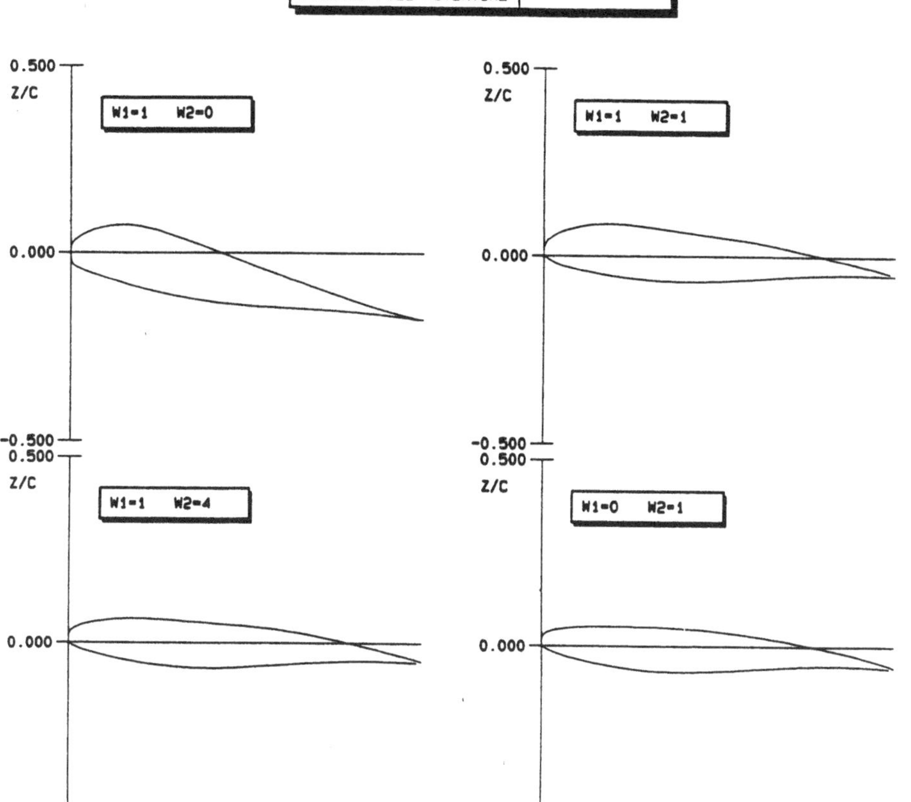

Fig. 10 Two-point airfoil design with different weight factor combinations; geometry

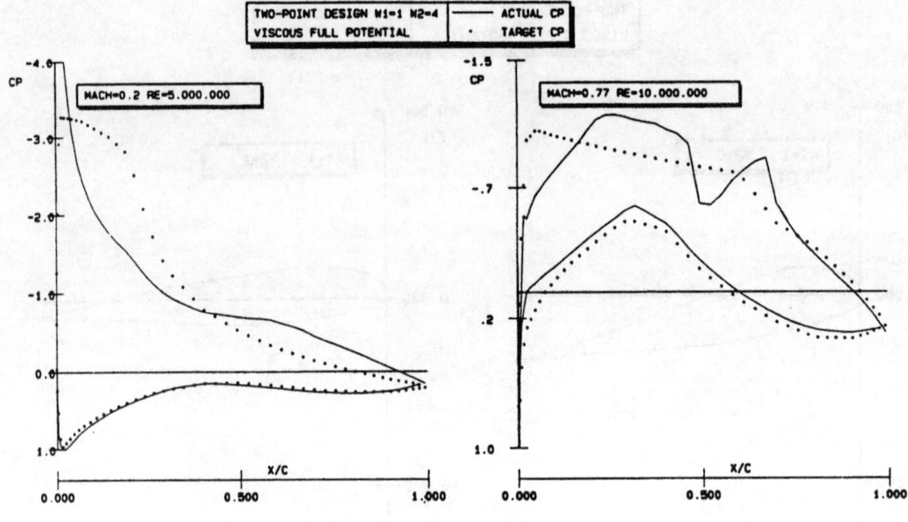

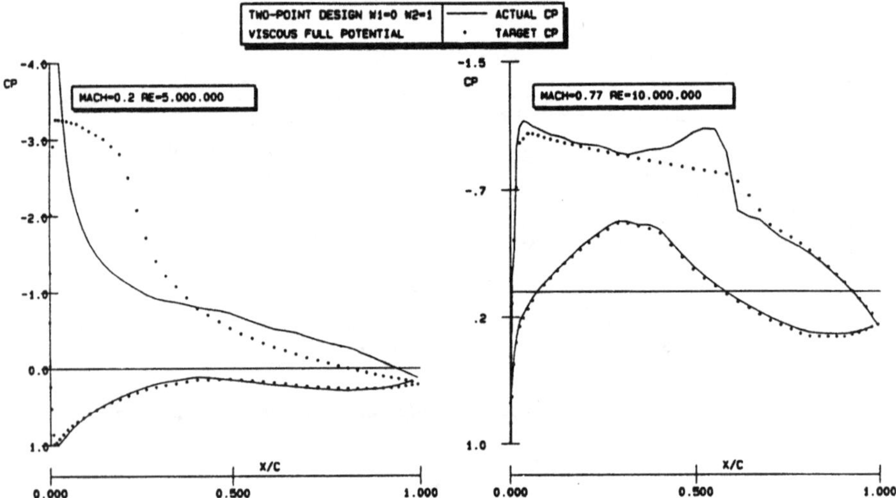

Fig. 11 (cntnd) Two-point airfoil design with different weight factor combinations; pressure distributions

by modifying a calculated pressure distribution, such that, at a certain lift coefficient, the supersonic pressure peaks were reduced to subsonic values, retaining approximately the original span lift distribution.

This design problem requires modifications at the wing nose region mainly. It is known to represent a not so easy case for the NLR approach, which relies on a linearized approximation of the inverse problem. The resulting wing geometry is shown for five computational sections in Fig.12 in comparison with the initial geometry. Comparison of the pressure distribution on the designed wing with the target pressure distribution at the same sections (see Fig.13) shows that it appears possible to design a wing geometry with the desired characteristics along almost the full span. Fig.14 shows the resulting isobars. The leading edge region shows fairly straight contours.

Fig.15 gives the values of the L_1 and L_2 norms defined by

$$L_1(C_p) = \frac{1}{N}\sum_{i=1}^{N} |C^i_{p_a} - C^i_{p_t}|$$

$$L_2(C_p) = \frac{1}{N}\sqrt{\sum_{i=1}^{N} (C^i_{p_a} - C^i_{p_t})^2}$$

as a function of the number of iterations (analysis computations). These figures, showing a sizeable reduction of the norms in iteration 0 to 13, are typical for straightforward application of the residual-correction approach starting with a given estimate of the wing geometry. However, at that point of computation, the design process was continued with refined adjustments of the geometry and geometry constraints in order to improve geometry regularity (both chordwise and spanwise). This "fine tuning" requires a relatively large effort, but is quite effective in geometry terms and results in an improved approximation of the target pressure distribution. The obtained reduction of the norms is about 1 to 2 orders of magnitude.

CONCLUDING REMARKS

Results have been presented of an iterative method to solve the multi-point airfoil design problem for subsonic and transonic conditions. The method minimizes a cost function, which weighs the deviations from specified target pressure distributions in a least squares sense.

The method is based on a residual correction approach and assumes the availability of a computer program for the analysis of the full-potential flow around an airfoil, possibly taking boundary layer effects into account. Deviations from the specified target pressure distributions (the residuals) are translated into airfoil geometry corrections, using an approximate inverse method based on an equivalent incompressible flow problem for each design point. This is in favour of computational speed, because fast integral equation methodology (panel methods) is then at disposal. The approximate inverse method is formulated as an equivalent incompressible multi-point airfoil design problem.

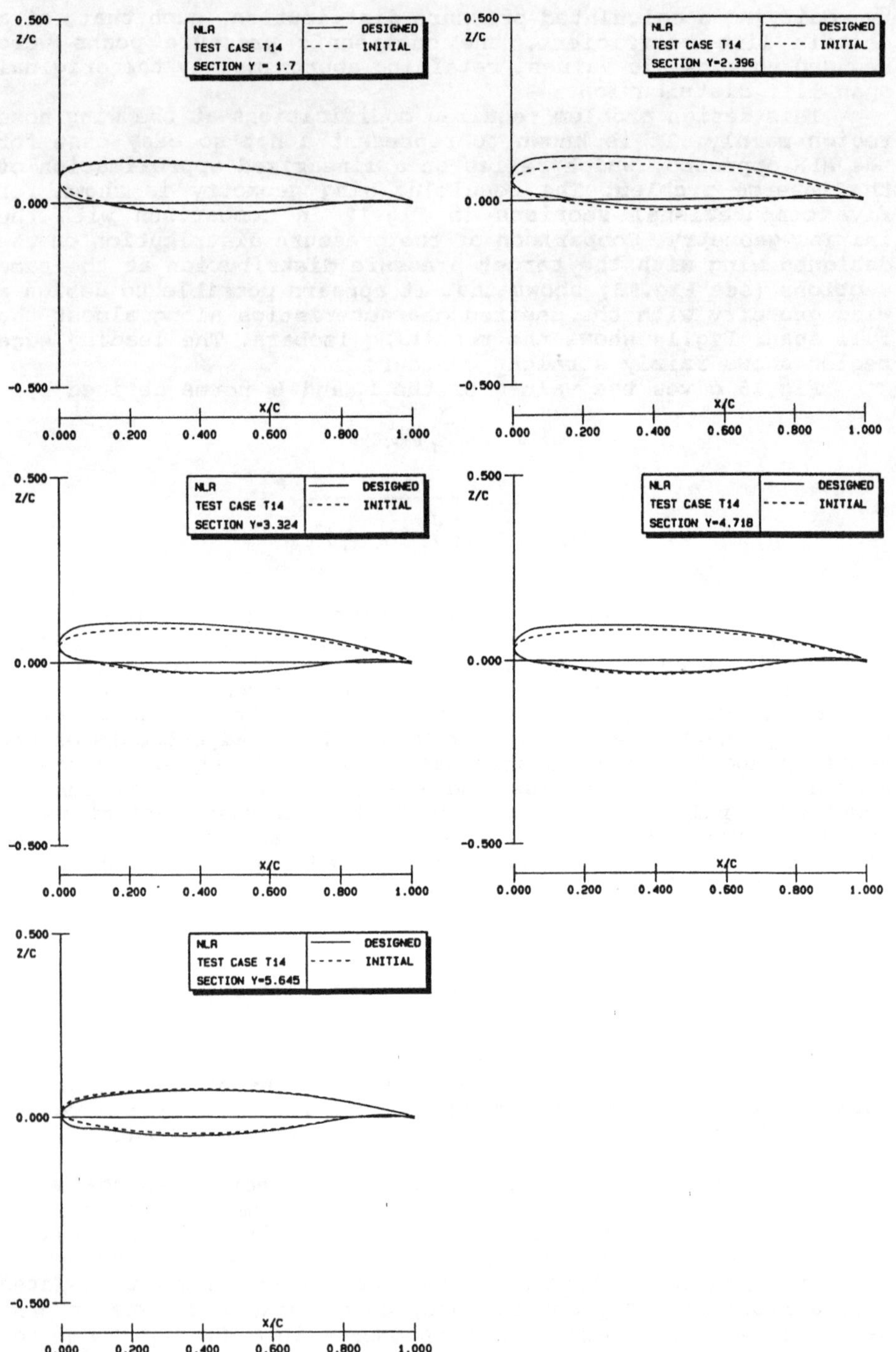

Fig. 12 Wing design: initial and resulting geometry

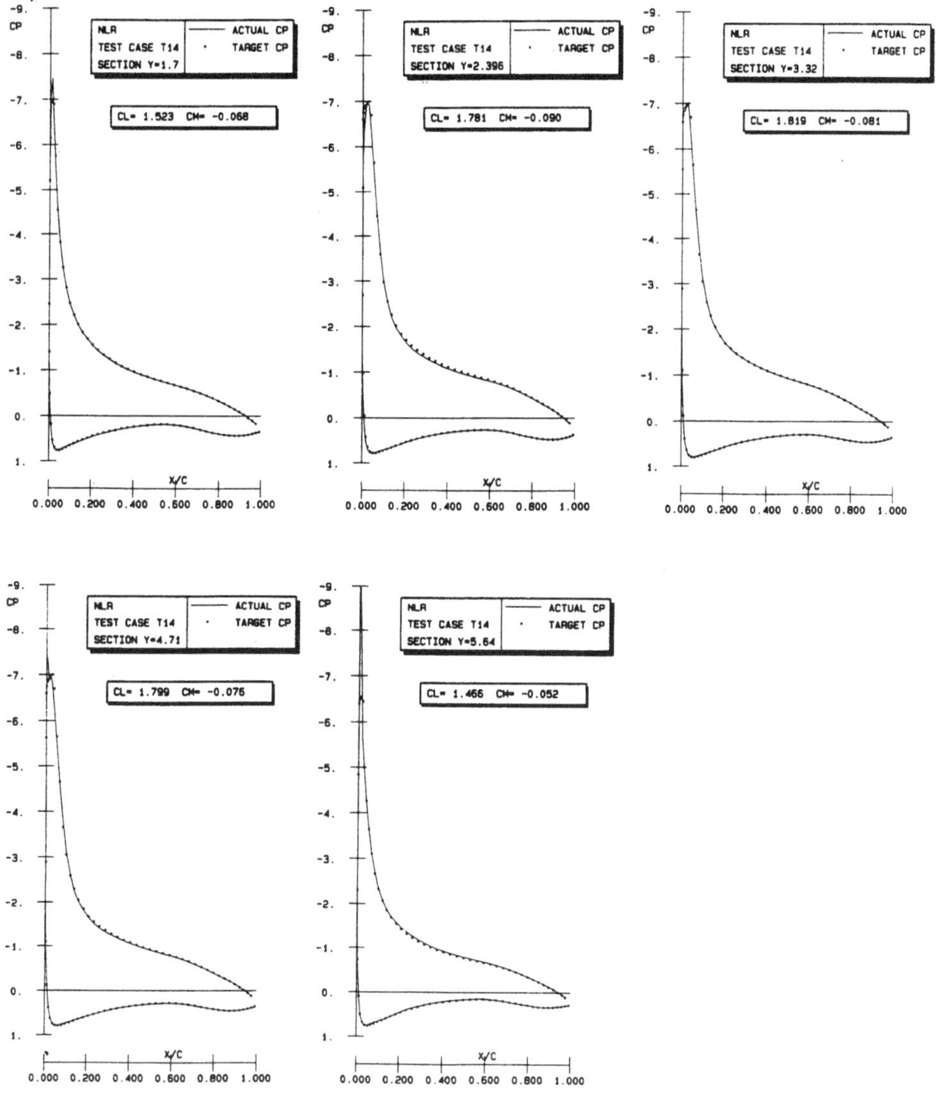

Fig. 13 Wing design: chordwise pressure distributions

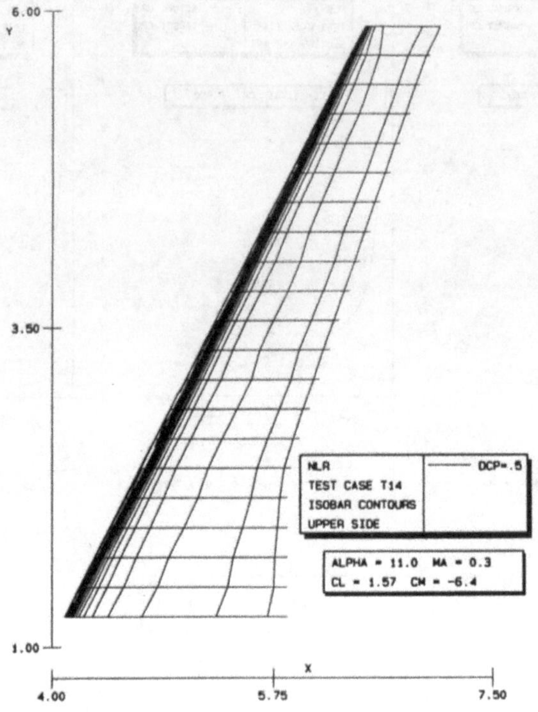

Fig. 14 Wing design: isobar pattern

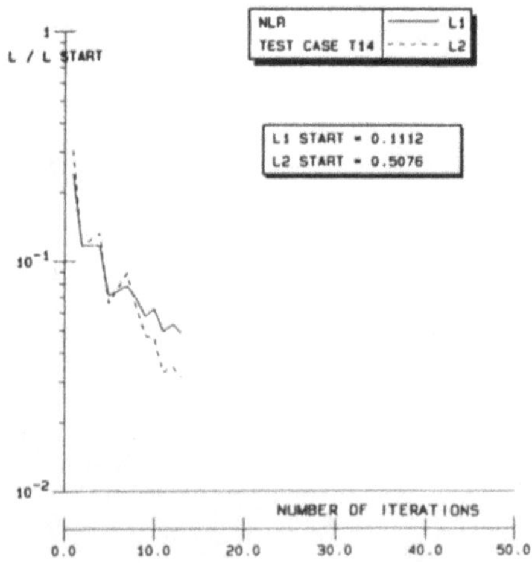

Fig. 15 Wing design: convergence history of wing pressure distribution

The feasibility of the introduction of the equivalent incompressible multi-point design problem has been demonstrated for a two-point reconstruction test case using a least squares approach for the solution of the linearized inverse equivalent incompressible flow problem. Likewise, a two-point example design case has underlined the point of view that multi-point design must be addressed as a re-design problem. Applicability of the residual-correction approach to 3D wing design has been demonstrated by means of a single-point problem.

Though the multi-point airfoil design method is formally based on transonic full-potential flow, its applicability is not strictly limited to such flow. In view of the residual correction approach, which only requires geometry corrections in the right direction, extension to inviscid flow governed by the Euler equations, or even to viscous flow, is straightforward, as long as a computer program for the analysis of such flow is available and the flow conditions considered are such that transonic full potential flow is a reasonable approximation of the inviscid outer flow. This points out the flexibility of the method.

REFERENCES

[1] Lighthill, M.J.: "A new method of two-dimensional aerodynamic design" R & M 2112, 1945

[2] Timman, R.: "The direct and the inverse problem of airfoil theory. A method to obtain numerical solutions",NLL F.16,-1951

[3] Drela, M.: "Newton solution of coupled viscous/inviscid multi-element airfoil flows , AIAA-90-1470, 1990

[4] Labrujère, Th. E., van der Vooren, J.: "Development of an algorithm for computational multi-point wing design; Algorithm definition", NLR report to be published.

[5] Fray, J.M.J., Slooff, J.W., Boerstoel, J.W., Kassies, A. "Design of transonic airfoils with given pressure subject to geometric constraints" NLR TR 84064 U, 1984.

[6] Brandsma, F.J., Fray, J.M.J.: "A system for transonic wing design with geometric constraints based on an inverse method." NLR TP 89179 L, 1989.

COMPARISON OF SOME OPTIMIZATION METHODS AND THEIR APPLICATION IN THE SHAPE OPTIMIZATION OF LIFTING AIRFOILS

O. Pironneau – A. Vossinis
INRIA – Rocquencourt
B.P. 105, 78153 Le Chesnay Cedex – FRANCE

SUMMARY

We propose the use of nonlinear GMRES in order to solve the optimality conditions in unconstrained optimization problems. We show academic examples and industrial applications of shape optimization in potential flows.

INTRODUCTION

Nonlinear GMRES is a robust algorithm for solving systems of equations fast. Therefore, its use in Shape Optimization in Fluid Mechanics is attractive. We suggest two ways of use: either we solve by GMRES the optimality conditions or we minimize the cost function projected on the Krylov subspace used by GMRES.

A comparison of the above mentioned ways of use with some other optimization algorithms (namely: Steepest Descent, Powell's Method and Least Squares Method) shows that GMRES for solving the optimality conditions (grad-GMRES) works best. The comparison is done with the help of an academic example of shape optimization in incompressible potential flow. We show also a simple application of airfoil shape optimization in the same kind of flow.

Then, we use grad-GMRES in two applications more industrial and less academic:

• Pressure recovery inverse problem for an airfoil in subsonic potential flow. The gradient calculation of the cost function is done by the use of an adjoint state ("exact computation" of the gradient).

• Shape Optimization of airfoils in transonic potential flow using an "exact computation" of the gradient of the cost function. We wish either to achieve a certain pressure distribution on the airfoil or to reduce the shock induced drag.

COMPARISON OF OPTIMIZATION METHODS

An optimization problem

We compare some optimization algorithms in the following shape optimization problem

Find the nozzle which realizes a certain velocity distribution on its walls for an incompressile, two-dimensional, irrotational and steady flow.

The flow is modeled by the use of a stream function Ψ, therefore the state equation is the following

$$\left.\begin{array}{c} -\Delta\Psi = 0 \text{ in } \Omega \\ \dfrac{\partial\Psi}{\partial\vec{n}}\,|_{\Gamma_1\cup\Gamma_3} = 0 \\ \Psi\,|_{\Gamma_2} = 1 \\ \Psi\,|_{\Gamma_4} = 0 \end{array}\right\} \tag{1}$$

where we assume that the flow velocity is parallel to the nozzle axis and uniform at the entrance and the exit of the nozzle and the wall and axis of the nozzle are streamlines (see Figure 1). P^1 finite elements are used for solving the state equation. The subscript h denotes the discretization of the geometry and of the solution.

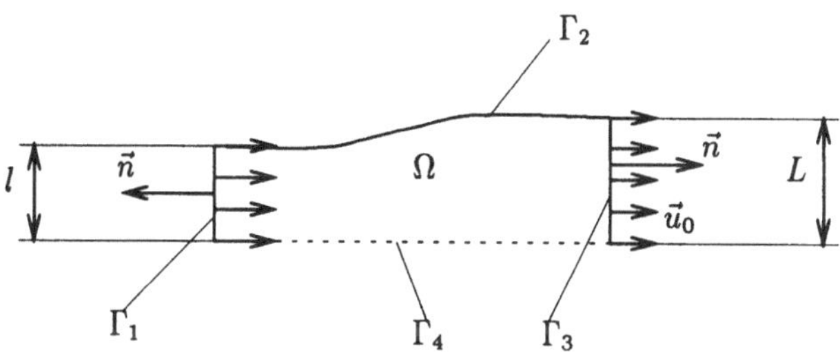

Figure 1: Notation on the nozzle section

The cost function is equal to

$$E(\Gamma_2) = \int_{\Gamma_2} |\,\nabla\Psi - \vec{c}\,|^2\, d\Gamma, \tag{2}$$

with $\vec{v} = (v_1, v_2)$ the desired velocity distribution on Γ_2 and $\vec{c} = (-v_2, v_1)$.

The gradient of the cost function with respect to the mesh point coordinates (discrete gradient) is equal to

$$\begin{aligned}
\frac{\partial E_h}{\partial q_{kl}} &= \int_{\Omega_h} [(\nabla w_k.\nabla P_h)\frac{\partial \Psi_h}{\partial x_l}(q_k) \\
&+ (\nabla \Psi_h.\nabla w_k)\frac{\partial P_h}{\partial x_l}(q_k) - (\nabla \Psi_h.\nabla P_h)\frac{\partial w_k}{\partial x_l}]dx \\
&- 2\int_{A_h} (\nabla \Psi_h - \vec{c}).\nabla w_k \frac{\partial \Psi_h}{\partial x_l}(q_k)dx \\
&+ \int_{A_h} |\nabla \Psi_h - \vec{c}\,|^2 \frac{\partial w_k}{\partial x_l}dx,
\end{aligned} \tag{3}$$

where

A_h the set of elements with one edge on the wall,
q_{kl} the l coordinate of the k mesh point,
w_k the shape function associated to the k mesh point,
P_h the solution of the adjoint equation:
Find $P_h \in V_{0h}$ such that

$$\int_{\Omega_h} \nabla P_h \cdot \nabla w_h dx = 2\int_{A_h} (\nabla \Psi_h - \vec{c}) \cdot \nabla w_h dx, \forall w_h \in V_{0h} \tag{4}$$

and

$$V_{0h}(\Omega_h) = \{w_h \in C^0(\Omega_h) : w_h\,|_{T_j} \in P^1, w_h = 0 \text{ on } \Gamma_{2h} \cup \Gamma_{4h}\}.$$

The optimization algorithms

Algorithms which need the calculation of the functional gradient

a) Steepest descent method

In order to find

$$\min_{z \in R^n} E(z)$$

we use a descent algorithm on the direction of $gradE$ which involves a search of the optimal descent step in each iteration (see [6]).

b) Generalized Minimal Residual Algorithm (GMRES)

We use nonlinear GMRES (cf [7]) in order to solve the optimality condition ($gradE = 0$) in an optimization problem with no constraints.

Algorithms which do not need the calculation of the functional gradient

a) GMRES as a direct minimization tool

Nonlinear GMRES solves a system of equations $F(z) = 0$ by minimizing the norm square of the residual projected on a Krylov subspace. Since the cost function is a sum of squares, we can construct a vector $F(z)$ whose norm square is equal to the cost function, then use nonlinear GMRES in order to solve $F(z) = 0$ and therefore minimize $|F|^2$.

b) Powell's method

The outline of Powell's method for minimizing a function $E: R^n \longmapsto R$ is the following:

- Choose a point $z \in R^n$ and a direction $x \in R^n$
- Find $\min_{\lambda \in R} E(z + \lambda x)$ and set $z = z + \lambda x$
- Choose a new direction $x \in R^n$ and solve the previous minimization problem.

c) Least Squares Method

The remark that E is a sum of squares naturally incites to minimize it by the use of Least Squares Method. We have used for that a routine of Harwell Computer Library which contains the Least Squares Method applied to the linearization of E.

Numerical results

We have tested the optimization methods in two kinds of problems:

a) Inverse Problem

The given velocity distribution is produced by a known nozzle. In this case $minE = 0$. We are interested in the convergence speed of each method and its capability to find the solution.

b) Real Problem

The given velocity distribution is arbitrary and we cannot assert that $minE = 0$. We are interested in the convergence speed of the different methods and their capability to find a solution.

In both kinds of problems we plot the convergence of the cost function versus the number of Laplacian solutions (Figures 3,4) and the computing cost versus the number of optimization parameters (Figures 5,6).

We can see that GMRES used to solve the optimality condition gives satisfactory results in both kinds of problems. More precisely, when it is used in the real problem optimization works better than the other methods. This is due to the fact that GMRES is a robust method able to solve a nonlinear system fast. Therefore, we shall use it in the sequel in the shape optimization of lifting bodies.

SHAPE OPTIMIZATION OF AIRFOILS IN INCOMPRESSIBLE FLOW

Problem formulation

We present an academic example of shape optimization

Find a lifting airfoil that achieves a certain velocity distribution on its boundary in an incompressible, twodimensional, irrotational and steady flow.

The flow is modelled in terms of a potential ϕ such that $\vec{u} = \nabla\phi$ in the bounded domain Ω (see Figure 2) which approximates the unbounded flow domain. The velocity of the unperturbed flow is uniform and the flow slips on the airfoil. In order to calculate the lift, we consider a potential discontinuity on a cut joining the trailing edge P with the downstream far field boundary; we calculate the potential jump by the use of Joukowski's condition.

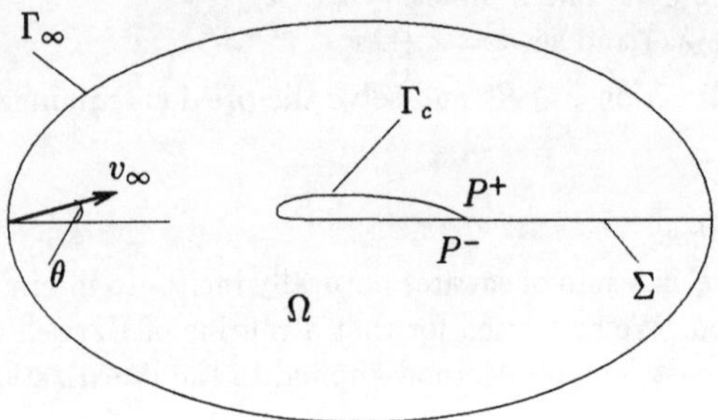

Figure 2: Computational domain around an airfoil

So, the state equation is

$$\left.\begin{array}{l} -\Delta\phi = 0 \text{ in } \Omega - \Sigma \\ \dfrac{\partial\phi}{\partial n} \mid_{\Gamma_\infty} = \vec{u}_\infty \cdot \vec{n} \\ \dfrac{\partial\phi}{\partial n} \mid_{\Gamma_c} = 0 \\ \phi \mid_{\Sigma^+} - \phi \mid_{\Sigma^-} = \alpha \\ \mid \nabla\phi(P^+) \mid^2 = \mid \nabla\phi(P^-) \mid^2 \\ \phi(P) = 0 \end{array}\right\} \quad (5)$$

In order to solve the state equation, we split the potential (see [3], [1]) $\phi = \phi_0 + \alpha\phi_1$ in a continuous nonlifting part ϕ_0 and in a lifting discontinuous part ϕ_1 with discontinuity equal to 1 and we use a P^1 finite element weak formulation.

Problem solution

The airfoil is discretized by a cubic spline. The optimization constraints are to keep fixed the airfoil length and a minimum airfoil surface. A first choice of optimization parameters would be the ordinates of the spline control points, which allows to keep a constant airfoil length. However, such a choice still needs the use of a constrained optimization method in order to take into account the second constraint. In order to use an unconstrained optimization technique, we do the following transforms: suppose that the ordinate $v_i \in [a_i, b_i]$. We map the interval $[a_i, b_i]$ to $[-1, 1]$, so v_i is mapped to $t_i = 2\dfrac{(v_i - a_i)}{(b_i - a_i)} - 1$. Then we optimize with respect to $z_i = arccos(t_i)$ with no constraints.

We solve the optimization problem by the use of grad-GMRES. The discrete gradient has an expression similar to eq. (3) when a velocity recovery criterion is used.

Actually, we need the functional gradient with respect to the optimization parameters. So, by taking into account the derivatives of the above mentioned transforms, we have

$$\frac{\partial E_h}{\partial z_i} = \frac{\partial E_h}{\partial v_i} \cdot \frac{\partial v_i}{\partial t_i} \cdot \frac{\partial t_i}{\partial z_i} \tag{6}$$

with $\frac{\partial v_i}{\partial t_i} = \frac{1}{2}(b_i - a_i)$ and $\frac{\partial t_i}{\partial z_i} = -sin(t_i)$, when $v_i \in [a_i, b_i]$.

Numerical experiments and results

a) Tests with the velocity recovery criterion

We present an inverse problem ($minE = 0$) at which we want to recover the velocity distribution of Korn's airfoil having as starting airfoil the NACA64A410.

b) Tests with the pressure recovery criterion

The same inverse problem with a pressure recovery criterion is also presented.

Figure 7 shows the convergence history of the cost functional for the two inverse problems. Figure 8 shows the comparison of the final airfoil and Korn's airfoil (this figure is identical for the two test cases).

SHAPE OPTIMIZATION OF AIRFOILS IN COMPRESSIBLE SUBSONIC FLOW

We present the solution of the following optimization problem

Find a lifting airfoil that achieves a certain pressure coefficient (C_p) distribution on its surface in a subsonic, two-dimensional, irrotational and steady flow.

The state equation

The flow is modeled again with the help of a potential Φ whose gradient is scaled by the critical speed of sound c_*, so the flow velocity is equal to $\vec{u} = \nabla\phi = \frac{\nabla\Phi}{c_*}$. We keep the notation of the incompressible problem. The state equation is

$$\nabla \cdot (\rho\nabla\phi) = 0 \text{ in } \Omega, \tag{7}$$

$$\rho\frac{\partial \phi}{\partial \vec{n}} \mid_{\Gamma_c} = 0, \tag{8}$$

$$\rho\frac{\partial \phi}{\partial \vec{n}} \mid_{\Gamma_\infty} = \rho\frac{\vec{v}_\infty \cdot \vec{n}}{c_*} = \rho_\infty g, \tag{9}$$

$$\phi(P) = 0, \tag{10}$$

$$\phi \mid_{\Sigma^+} - \phi \mid_{\Sigma^-} = \alpha, \tag{11}$$

$$\mid \nabla\phi(P^+) \mid^2 = \mid \nabla\phi(P^-) \mid^2, \tag{12}$$

where the density of the fluid is

$$\rho(\phi) = (1 - \kappa \mid \nabla\phi \mid^2)^r \tag{13}$$

with $\kappa = \dfrac{\gamma-1}{\gamma+1}$, $r = \dfrac{1}{\gamma-1}$, γ the ratio of specific heats (c_p/c_v) of the fluid. In order to solve the state equation, we use again a potential splitting: $\phi = \tilde{\phi} + \alpha\phi_L$ (see [1], [3]), where ϕ_L is an incompressible, lifting potential which verifies the system

$$\left.\begin{aligned} -\Delta\phi_L = 0 \text{ in } \Omega - \Sigma \\ \frac{\partial\phi_L}{\partial n} \mid_{\Gamma_c\cup\Gamma_\infty} = 0 \\ \phi_L \mid_{\Sigma^+} - \phi_L \mid_{\Sigma^-} = 1 \\ \phi_L(P) = 0 \end{aligned}\right\} \tag{14}$$

and $\tilde{\phi}$ is a compressible, nonlifting potential, such that ϕ and α satisfy the eq. (7)-(12).

We use P^1 finite elements for solving the above systems of equations.

Solution of the optimization problem

We use the airfoil discretization and the choice of the optimization parameters described in the incompressible problem.

We compute the gradient of the cost function in two steps:

a) with respect to the mesh point coordinates

b) with respect to the optimization parameters by using eq. (6).

In order to compute the gradient with respect to the mesh point coordinates, we define a functional (subscript h denotes again P^1 discretized quantities)

$$F_h(q_{kl}, \tilde{\phi}_h, \tilde{\phi}_{Lh}, \alpha) = E_h(z_i)$$

and we call $B_1 = 0$ the system verified by $\tilde{\phi}_h, \phi_h$ and α, $B_2 = 0$ the system verified by ϕ_{Lh} and $B_3 = 0$ Joukowski's condition. Then, we introduce the Lagrangian

$$\mathcal{L}(q_{kl}, \tilde{\phi}_h, \tilde{\phi}_{Lh}, \alpha, \lambda_m) = F_h + \sum_{m=1}^{3} \lambda_m^t B_m$$

and the adjoint eq. system

$$(\frac{\partial B_1}{\partial\tilde{\phi}_h})^t \cdot \lambda_1 + (\frac{\partial B_3}{\partial\tilde{\phi}_h})^t \cdot \lambda_3 = -\frac{\partial F_h}{\partial\tilde{\phi}_h}$$

$$(\frac{\partial B_1}{\partial\alpha})^t \cdot \lambda_1 + (\frac{\partial B_3}{\partial\alpha})^t \cdot \lambda_3 = -\frac{\partial F_h}{\partial\alpha}$$

$$(\frac{\partial B_1}{\partial\tilde{\phi}_{Lh}})^t \cdot \lambda_1 + (\frac{\partial B_2}{\partial\tilde{\phi}_{Lh}})^t \cdot \lambda_2 + (\frac{\partial B_3}{\partial\tilde{\phi}_{Lh}})^t \cdot \lambda_3 = -\frac{\partial F_h}{\partial\tilde{\phi}_{Lh}}.$$

The gradient of the cost function is

$$\frac{\partial E_h}{\partial q_{kl}} = \frac{\partial F_h}{\partial q_{kl}} + \sum_{m=1}^{3} \lambda_m^t \cdot (\frac{\partial B_m}{\partial q_{kl}}).$$

The derivatives $\dfrac{\partial F_h}{\partial\tilde{\phi}_h}, \dfrac{\partial F_h}{\partial\alpha}, \dfrac{\partial F_h}{\partial\tilde{\phi}_{Lh}}$ are calculated by finite differences. Formal derivative calculations are performed for $\dfrac{\partial B_m}{\partial\tilde{\phi}_{Lh}}$ and $\dfrac{\partial B_m}{\partial q_{kl}}, m = 1, 2, 3$.

Numerical experiments and results

We present an inverse pressure coefficient (C_p) recovery problem: to find the C_p distribution of Korn's airfoil when the far field Mach number is equal to .65 and the angle of attack equal to zero, having as initial airfoil the NACA64A410 in the same flow.

The cost function is equal to

$$E(z) = \frac{1}{2}\int_{\Gamma_c}(C_p - C_{pd})^2 dx$$

with C_{pd} a given pressure coefficient distribution.

We see in Figure 9 the error between the "exact" functional gradient and the gradient calculated by finite differences and in Figure 10 the convergence history of the cost funciton and its gradient. The airfoil comparison is as in Figure 8.

SHAPE OPTIMIZATION OF AIRFOILS IN COMPRESSIBLE TRANSONIC FLOW

We solve a C_p recovery optimization problem, which reads as the corresponding subsonic problem, the only difference being that now a transonic flow is considered. We also present a drag minimization problem in the same kind of flow.

The state equation

We use again the notation of Fig. 2. The flow quantities are found as the steady state solution of the mass continuity equation

$$\frac{\partial \rho}{\partial t} + \nabla(\rho \vec{u}) = 0. \tag{15}$$

A potential ϕ is introduced such that $\vec{u} = \vec{v}_\infty + \nabla\phi$. An expression for the density is found by using Bernoulli's equation:

$$\rho = [1 + \frac{\gamma - 1}{2} M_\infty^2 (1 - (2\frac{\partial \phi}{\partial t} + u^2))]^{\frac{1}{\gamma-1}}. \tag{16}$$

The boundary conditions on the far field and airfoil boundaries are the ones of the subsonic case. The lift is taken into account by introducing a potential jump on the cut Σ (see Fig. 2) and the pressure continuity through the cut is ensured by the use of an approximated Riemann solver adapted in potential flows (cf [5],[4]). The same Riemann solver is also used in order to treat far field boundary conditions.

A P^1 finite element formulation and a first order time stepping scheme is used in order to solve the state equation. An mass flux upwinding is added to the weak formulation of the state equation in order to capture supersonic pockets and to avoid nonphysical shocks which violate the entropy condition.

Solution of the problem

We use again the airfoil discretization and the choice of optimization parameters of the incompressible problem. The optimization method is grad-GMRES and the discrete functional gradient is calculated in a formal way with the help of the Lagrangian, treating the state equation as a constraint and introducing the corresponding adjoint state equation (cf [2]).

Numerical experiments and results

a) Pressure recovery inverse problem

The cost function is the same as in the subsonic case. The test presented is to recover C_p distribution of Korn's airfoil in a flow with $M_\infty = .75$ and zero angle of attack having as initial airfoil the NACA64A410. Figure 11 shows the relative error between the "exact" gradient and the gradient computed by finite differences. Figure 12 shows the convergence history of the functional value and its gradient. The airfoil comparison is as in Figure 8.

b) Drag minimization problem

We wish to find a shock free airfoil in a flow with $M_\infty = .73$ and angle of attack equal to 2°, keeping the lift the closest possible to its initial value. Therefore, we introduce the cost function

$$E(z) = \frac{1}{2}\int_{\Gamma_c}(C_p - C_{pinit})^2 dx + \beta C_D,$$

where C_{pinit} is the C_p distribution of the initial airfoil, C_D the shock induced drag coefficient and β a weighting factor whose value depends on the trade off between drag reduction and lift conservation.

We present at the Figures 13 – 20 the results of a test case when the initial airfoil is the RAE2822.

REFERENCES

[1] M.O. Bristeau. Application of a Finite Element Method to Transonic Flow Problems using an Optimal Control Approach. Technical report, Von Karman Institute for Fluid Dynamics, 1978.

[2] J. Cea. Conception Optimale ou Identification de Formes: calcul rapide de la dérivée directionnelle de la fonction coût. Université de Nice, 1984.

[3] D.H. Norries – G. Devries. *The Finite Element Method.* Academic Press, New York, 1973.

[4] F. Dubois. Conditions aux limites fortement non-linéaires pour les équations d'Euler. Notes des cours de l'Ecole d'Eté CEA–EDF–INRIA sur les problèmes non-linéaires appliqués, 1988.

[5] F. Dubois. *Quelques problèmes liés au calcul d'écoulements de fluides parfaits dans les tuyères.* PhD thesis, Univ. Paris VI, 1989.

[6] O. Pironneau. *Optimal Shape Design for Elliptic Systems.* Springer Series in Computational Physics, New York, 1984.

[7] L.B. Wington - N.J. Yu - D.P. Young. GMRES Acceleration of Computational Fluid Dynamics Codes. *American Institute of Aeronautics and Astronautics*, 1984.

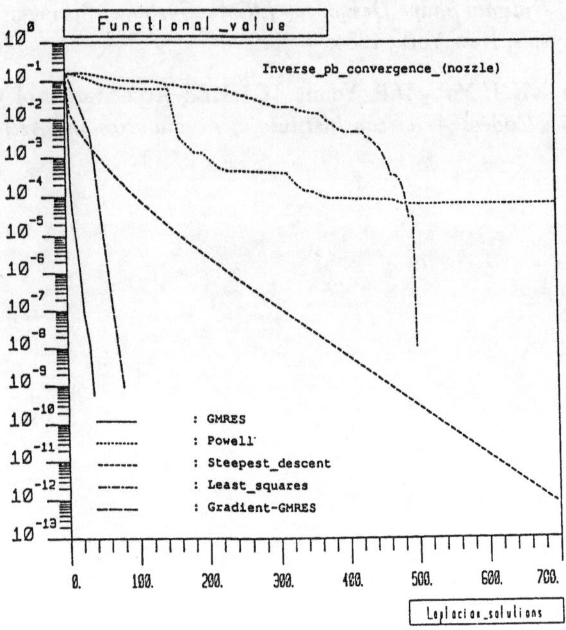

Figure 3: Nozzle optimization with 25 optimization parameters. Convergence comparison for an inverse problem

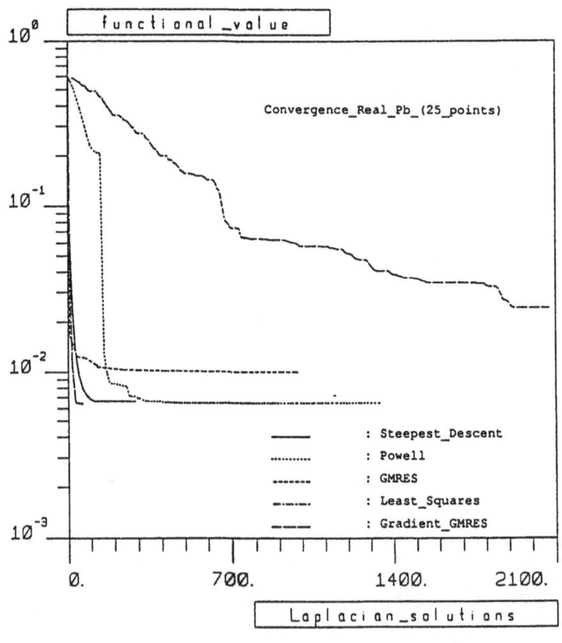

Figure 4: Nozzle optimization with 25 optimization parameters. Convergence comparison for a real problem

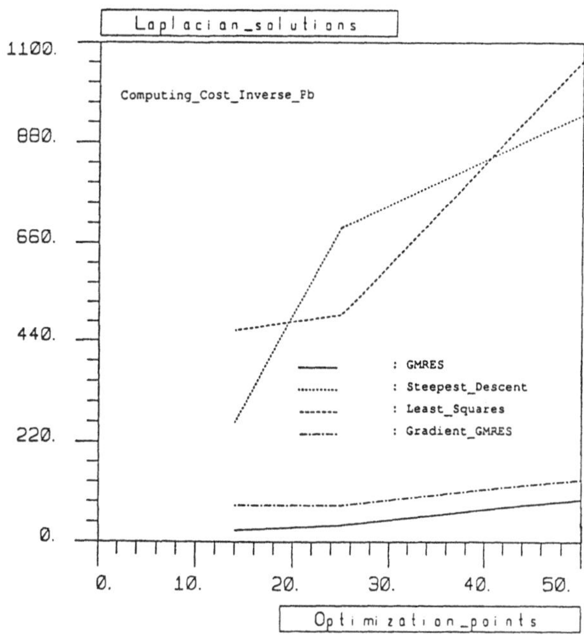

Figure 5: Computing cost vs the number of optimization parameters for an inverse problem

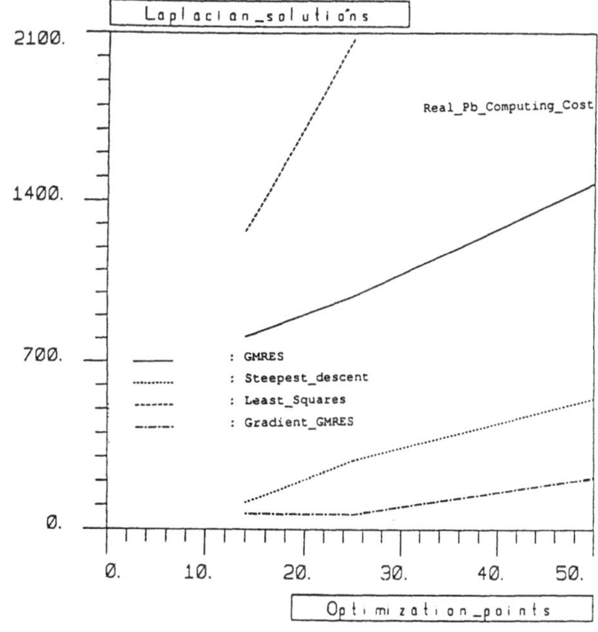

Figure 6: Computing cost vs the number of optimization parameters for a real problem

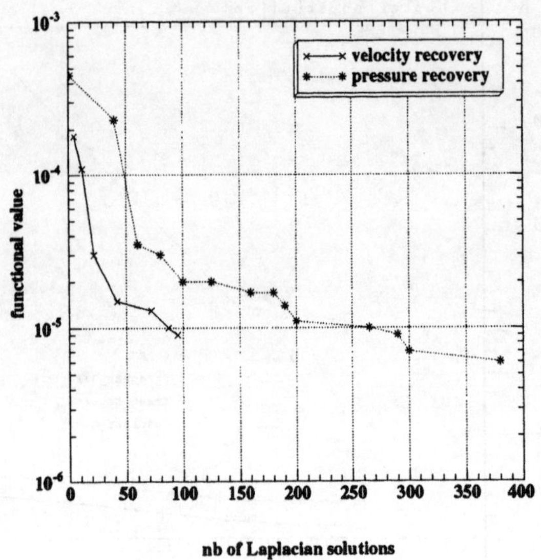

Figure 7: Convergence of Gradient-GMRES for inverse problems in incompressible flow. Initial airfoil NACA64A410, final Korn's airfoil. Optimization with 20 optimization parameters.

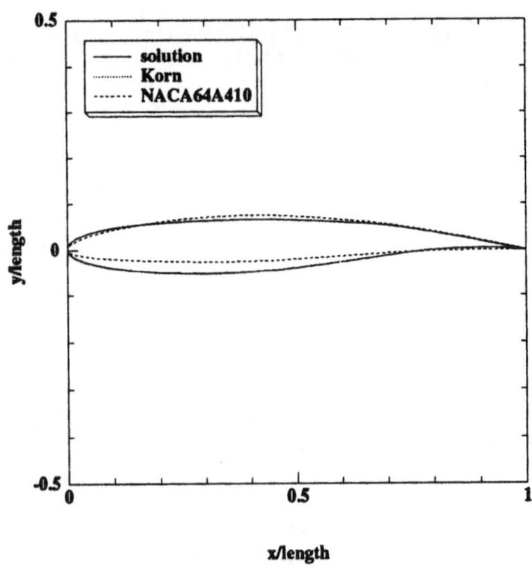

Figure 8: Airfoil convergence. Initial airfoil NACA64A410, final Korn's airfoil.

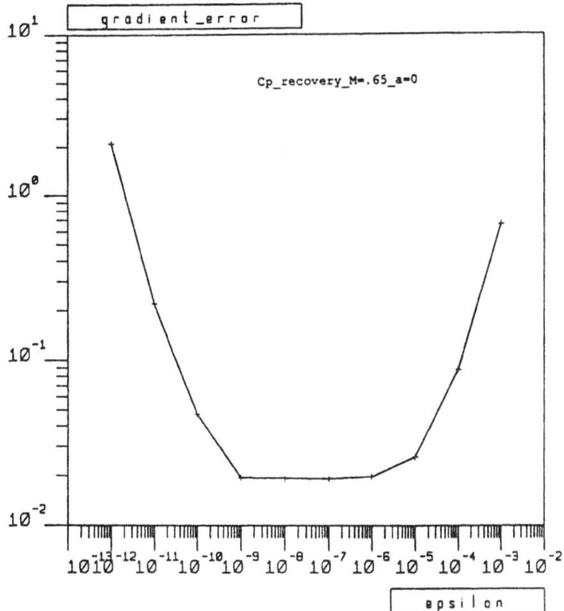

Figure 9: Cp recovery inverse problem, M_∞ = .65, angle of attack=0, initial airfoil NACA64A410, final Korn's airfoil. Exact and Finite Differences gradient error

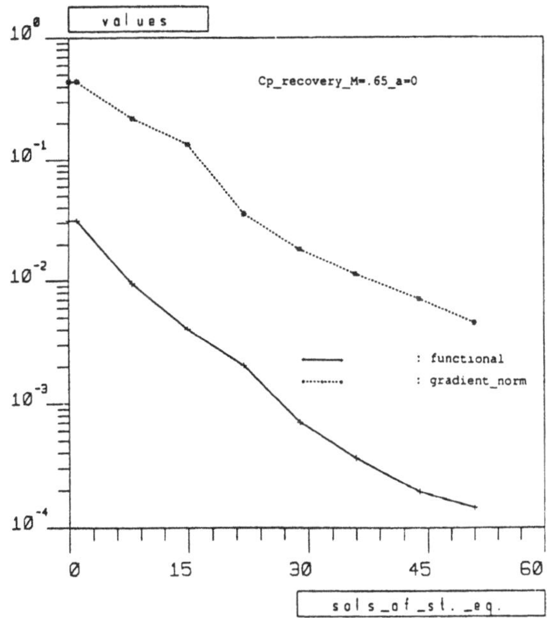

Figure 10: Cp recovery inverse problem, M_∞ = .65, angle of attack=0, initial airfoil NACA64A410, final Korn's airfoil. Functional and gradient norm convergence vs nb of functional calls

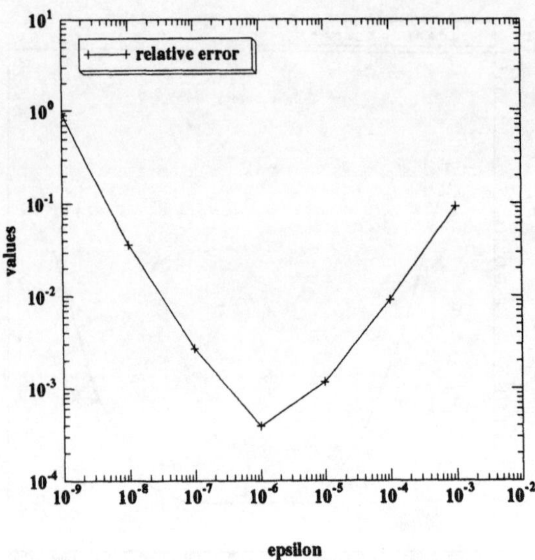

Figure 11: Cp recovery inverse problem, $M_\infty = .75$, angle of attack=0, initial airfoil NACA64A410, final Korn's airfoil. Relative error of functional gradient

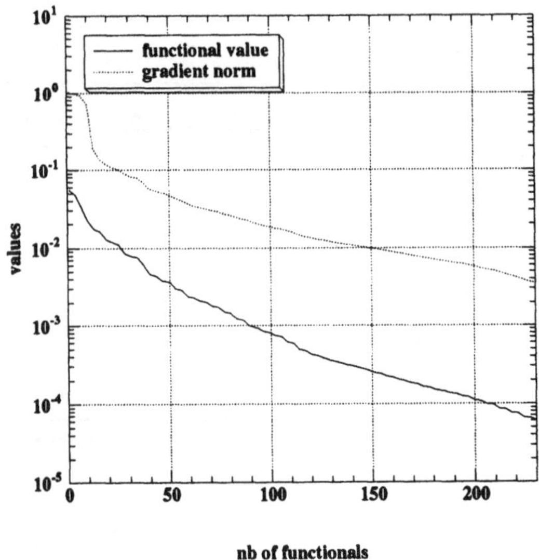

Figure 12: Cp recovery inverse problem, $M_\infty = .75$, angle of attack=0, initial airfoil NACA64A410, final Korn's airfoil. Functional and gradient convergence

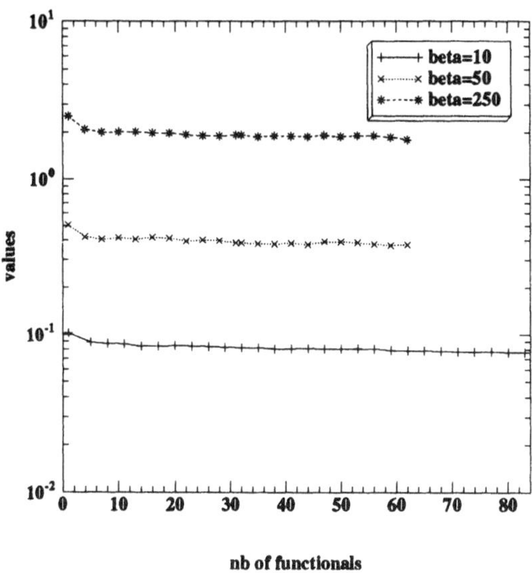

Figure 13: Drag minimization, $M_\infty = .73$, angle of attack= 2°, initial airfoil RAE2822. Functional convergence for different values of β

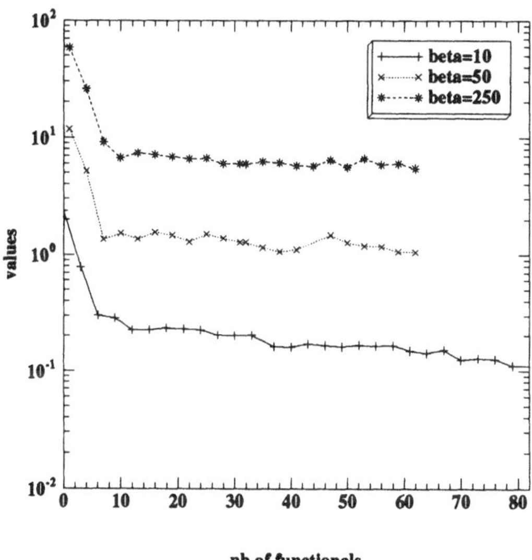

Figure 14: Drag minimization, $M_\infty = .73$, angle of attack= 2°, initial airfoil RAE2822. Gradient norm convergence for different values of β

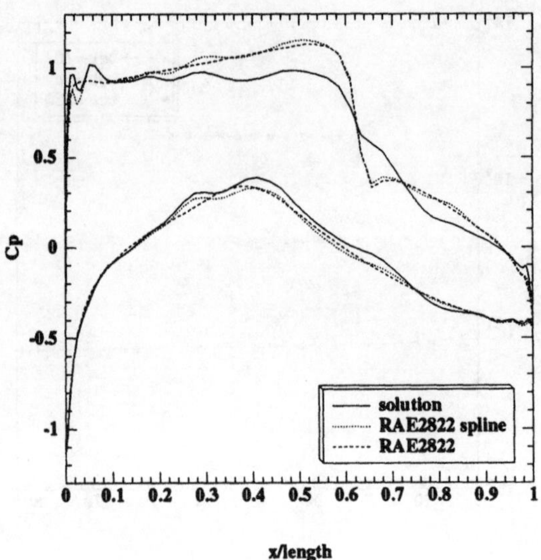

Figure 15: Drag minimization, M_∞ = .73, angle of attack= 2°, initial airfoil RAE2822. Initial and final Cp for beta = 10

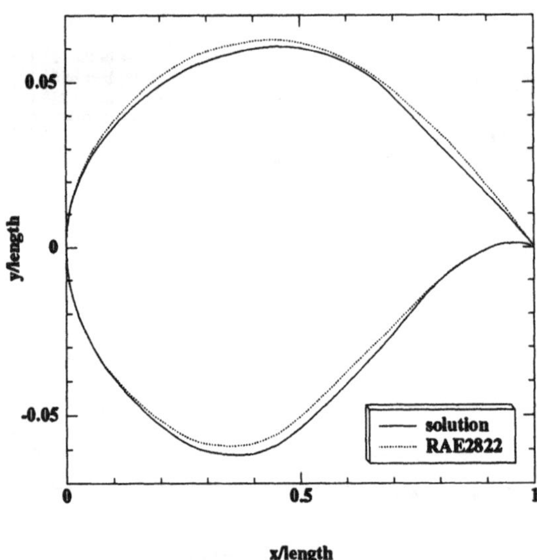

Figure 16: Drag minimization, M_∞ = .73, angle of attack= 2°, initial airfoil RAE2822. Initial and final airfoil for beta = 10

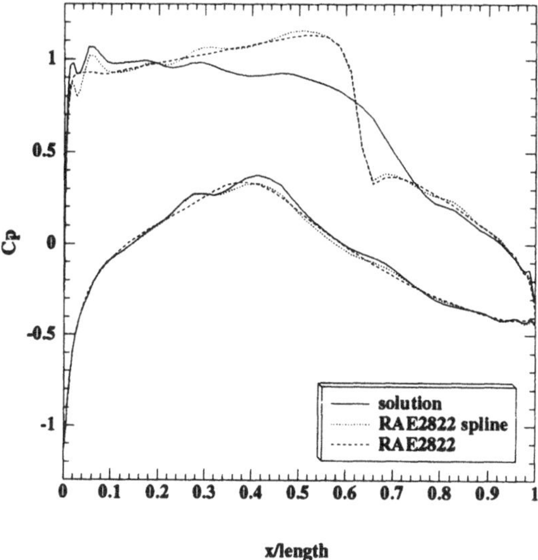

Figure 17: Drag minimization, $M_\infty = .73$, angle of attack= 2°, initial airfoil RAE2822. Initial and final Cp for beta = 50

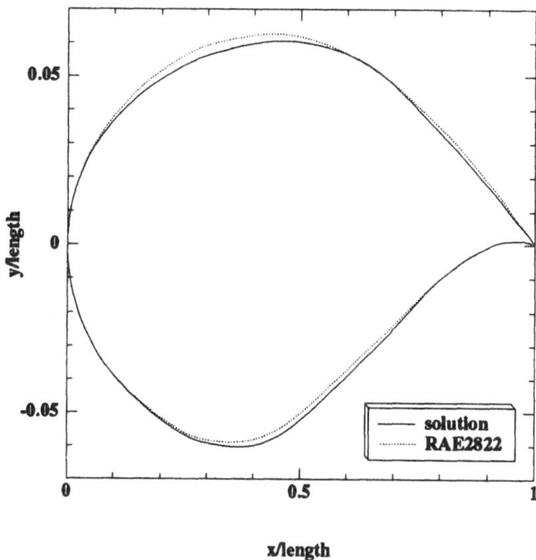

Figure 18: Drag minimization, $M_\infty = .73$, angle of attack= 2°, initial airfoil RAE2822. Initial and final airfoil for beta = 50

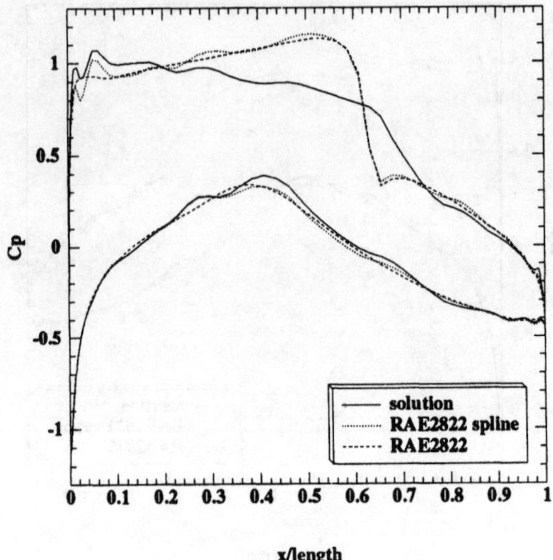

Figure 19: Drag minimization, $M_\infty = .73$, angle of attack= 2°, initial airfoil RAE2822. Initial and final Cp for beta = 250

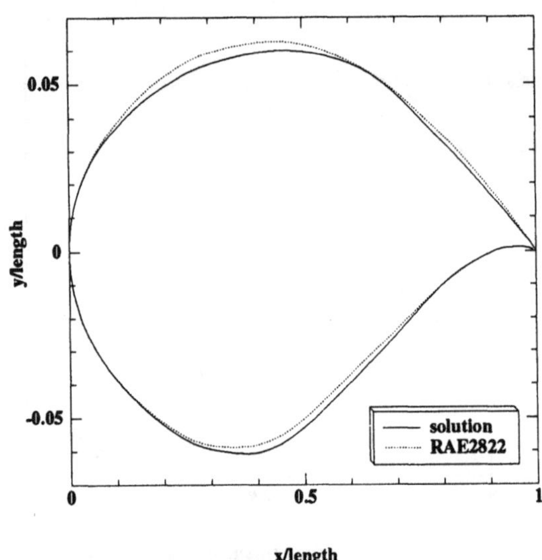

Figure 20: Drag minimization, $M_\infty = .73$, angle of attack= 2°, initial airfoil RAE2822. Initial and final airfoil for beta = 250

SHAPE OPTIMIZATION OF AN EULER FLOW IN A NOZZLE

François Beux

INRIA

B.P. 93, 06902 Sophia-Antipolis Cedex

France

SUMMARY

The optimization of an obstacle shape immersed in an Eulerian flow is investigated. The chain rule through the state equation is expressed by applying an adjoint state method to an upwind discretization with differentiable flux splitting. Thus, we construct a descent method with a discrete gradient computed exactly. In order to improve the performances, a method for increasing the number of unknowns, and relying on a multilevel idea is also implemented. These methods are applied to shape optimization of a 2-D nozzle in a subsonic or transonic Eulerian flow; inverse and optimization problems are successively considered.

INTRODUCTION

Shape Optimum Design in Aerodynamics has shown an important development in these last years; moreover, many numerical experiments have been performed, with models of increasing complexity, up to full potential flows. But the case of flows governed by Euler equations (proposed here) has not been extensively investigated. In fact, the main difficulty is related to the important cost of one single flow calculation for complex models. Moreover this cost must be multiplied by the points of design (several Mach regimes, ...) and the number of cost evaluations performed for optimization. We use an exact gradient algorithm which is more complex to construct and to program and less general (it requires a differentiable functional) than an approximate gradient algorithm; but we obtain a strong diminution of the number of cost evaluations. In fact, we have as many informations as number of optimization parameters for an evaluation of a gradient and of a cost function. In general, for complex models, only few optimization parameters are used; also we combine here the previous algorithm with a hierarchical strategy in which the number of shape parameters varies during the optimization procedure. A family of embedded parametrizations are built for describing smoothly the nozzle shape. The exact-gradient method for the Euler equations and the hierarchical optimization approach are described in detail respectively in [1] and [2].

METHODOLOGY

PROBLEM FORMULATION

We consider an optimum shape design problem which can be defined, in a general context, in the following way:
let Γ_{ad} be a family of shapes smooth enough; to any shape $\gamma \in \Gamma_{ad}$ we associate a domain Ω_γ and the solution $W(\gamma)$ of a partial derivative equation modelling the considered flow:

$$\forall \gamma \in \Gamma_{ad} \qquad \Psi(\gamma, W) = 0 \ .$$

From $W(\gamma)$ we construct a cost function associated with a given criterion:

$$j(\gamma) = J(\gamma, W(\gamma)) .$$

Finally we try to solve the following optimization problem :

$$\text{Find } \bar{\gamma} \in \Gamma_{ad} \text{ such that } \quad j(\bar{\gamma}) = \min_{\gamma \in \Gamma_{ad}} j(\gamma) \ .$$

In particular, we consider here a family of 2D nozzles; since they are symmetric, we can solve the problem for an half of the nozzle. We consider an ideal gas flow: the flow enters in the nozzle at $x = -2$ $(0 < y < .5)$ and exits at $x = 4$ $(0 < y < .5)$ The geometry of the nozzle that represents the boundary of Ω_γ is defined in Figure 1.
$W(\gamma)$ is solution of the steady Euler equations :

$$F(W(\gamma))_x + G(W(\gamma))_y = 0 \text{ in } \Omega_\gamma \ .$$

A slip boundary condition is applied at nozzle axis and wall.

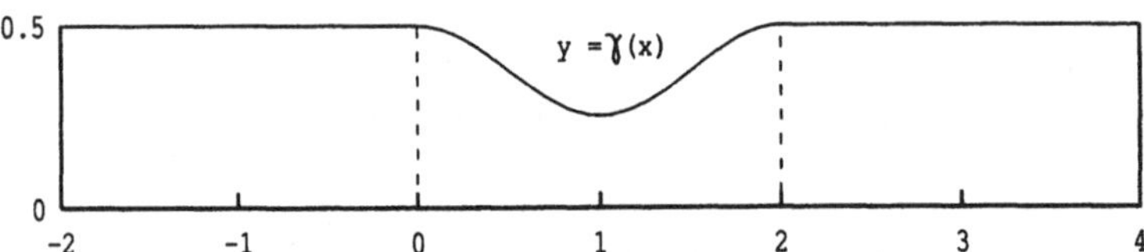

Figure 1 : Nozzle geometry

FLOW SOLVER AND NUMERICAL APPROXIMATION

The employed mesh is a triangular structured mesh (see Figure 2); the nodes are on the vertices. We use a triangular upwind finite volume formulation (see [6]), for each vertex i we have:

$$\Psi_{Euler}(W^n) = \sum_{j\epsilon\kappa(i)} \Phi(W_i^n, W_j^n, \int_{\partial C_{ij}} \vec{n}\, d\sigma) \quad + \quad \text{Boundary} \quad \text{conditions},$$

where $\kappa(i)$ is the set of neighboring nodes of i, ∂C_i the boundary of the polygonal ceil around node i, $\partial C_{ij} = \partial C_i \cap \partial C_j$, $\vec{n}$ the outward normal unit vector along ∂C_i, Φ a numerical flux function. The resulting approximation is first-order accurate.
We choose for Φ the Van Leer flux function which is differentiable [7]. Therefore for solving the discrete system approximating the steady Euler equations, we use an implicit linearized time advancing iteration :

$$(\frac{I}{\Delta t} + \Psi'(W^n))\, \delta W^{n+1} = -\Psi(W^n) \;, \tag{1}$$

where $\Psi'(W^n)$ is computed exactly. The linear system is solved by a collective (node-by-node) Jacobi relaxation.

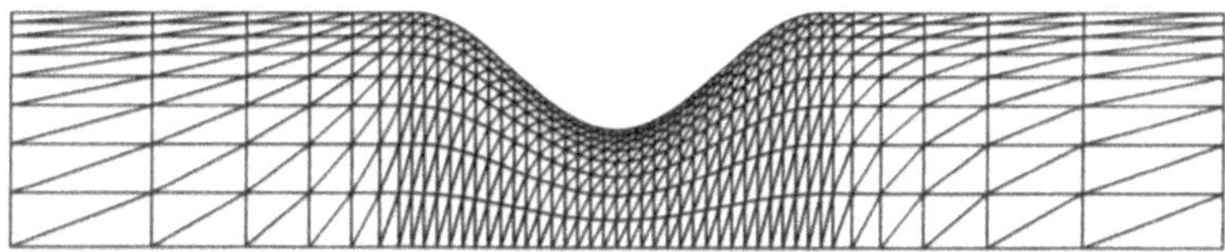

Figure 2 : Domain discretization

SHAPE PARAMETRIZATION AND MESH UPDATE

The nozzle mesh is a deformation of a rectangular mesh by a homothety in the y direction, for each straight vertical mesh line; the mesh coordinates are thus directly governed by the coordinates of the controlled upper boundary section (concertina-like mesh). The shape is controlled by the position of a selected number of its nodes; an interpolation maximizing the smoothness of the curve is then applied for finding the missing nodes.
We present here an example of interpolation: a nested cubic interpolation. We begin to interpolate at the middle of each $[x_i, x_{i+1}]$, using a cubic interpolation, we obtain :

$$(Pf)_j = \frac{9}{16}(f_i + f_{i+1}) - \frac{1}{16}(f_{i-1} + f_{i+2}) \;. \tag{2}$$

In Figure 3 the symbol " □ " represents the values of f_i of the first parametrization while the symbol " × " the interpolated values $(Pf)_j$; then, starting from the new set of points, we calculate by the same interpolation rule, an other set of points (symbolized in Figure 3 by " + "), and we repeat this procedure until all points are generated. $(Pf)_j$ is expressed by linear combination of $f_{i1}, \cdots, f_{il}$ values of f on nodes of the existing coarse level.

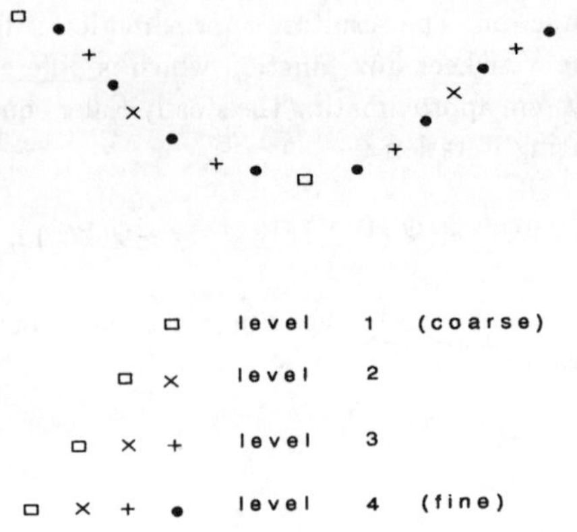

Figure 3 : Sketch of the nested interpolation of the shape

DERIVATION OF THE GRADIENT

The parameters of optimization are the ordinates of a subset of the boundary nodes, $\gamma = (\gamma_1, ..., \gamma_m)$. Using a concertina-like mesh defined previously, mesh's deformation results in an ordinate change and depends only on the control γ. So we can compute a Fréchet derivative with respect to the shape of $j(\gamma) = J(\gamma, W(\gamma))$. We obtain then:

$$\forall\, \delta\gamma \in I\!R^m \qquad < \frac{\partial j}{\partial \gamma}, \delta\gamma > = < \frac{\partial J}{\partial \gamma}, \delta\gamma > - < \Pi, \frac{\partial \Psi}{\partial \gamma}\delta\gamma > ,$$

with Π , the adjoint state, defined by the following linear system:

$$A_n \Pi = \frac{\partial J}{\partial W} \quad \text{with} \quad A_n = (\frac{\partial \Psi}{\partial W})^* .$$

The matrix $\frac{\partial \Psi}{\partial \gamma}$ is computed simultaneously with the flux Ψ, i.e. when $W(\gamma)$ is computed. However, $\Psi'(W^n)$ has already been computed in solving the Euler equations, by the unsteady implicit linearized time stepping. Hence, A_n can be deduced directly from Equation (1) without any additional calculation.

OPTIMIZATION ALGORITHM

The basic idea is to compute a descent direction with alternately different levels of parametrization.
At first, we consider the Full V-cycle strategy (according to the multigrid theory [4],[5]), which consists of the following phases :
we start doing several iterations on the coarsest level, then, in a second phase the optimization is made alternately on coarsest level and on next level, etc. In Figure 4 this strategy is illustrated for three levels.
We can also use a "nested iteration" strategy : beginning on the coarsest level, we converge successively on each level until the finest one (see Figure 5).

We summarize here the algorithm giving a descent direction at the nth optimization iteration :

- first, we have computed a gradient $g_n = grad\, j(u_n)$.
- Assuming that we are on level i (with l_i parameters); we can associate to this level an interpolation $P_{i,n} : \mathbb{R}^{l_i} \rightarrow \mathbb{R}^m$ defined previously (m is the total number of parameters). $g_n^F = P_{i,n}^* g_n$ is then a descent direction in the space $\mathbb{R}^{l_i}$.
- Finally, we obtain a descent direction on the finest level given by the following expression:

$$h_n = P_{i,n} P_{i,n}^* g_n .$$

- A Polak-Ribière conjugate gradient algorithm can be optionally employed, combined with the hierarchical strategy.
- We apply a 1-D minimization consisting of firstly a dichotomic research, and finally a parabolic interpolation that ensures a quadratic final convergence.

In order to summarize, a sketch of the gradient derivation and of the optimization algorithm is shown in Figure 6.

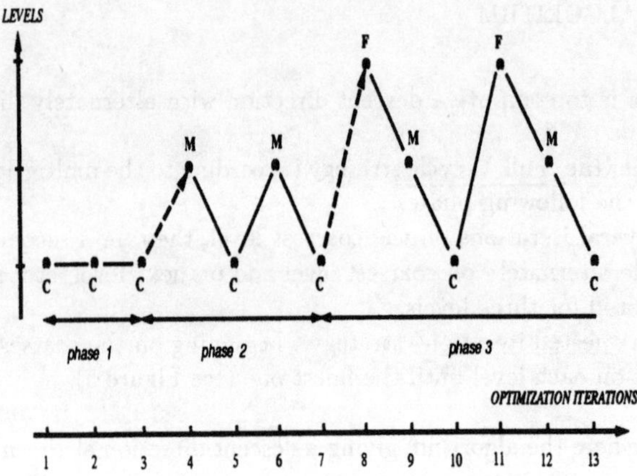

Figure 4 : Sketch of full V-cycle strategy on 3 levels (C for coarse, M for medium, F for fine)

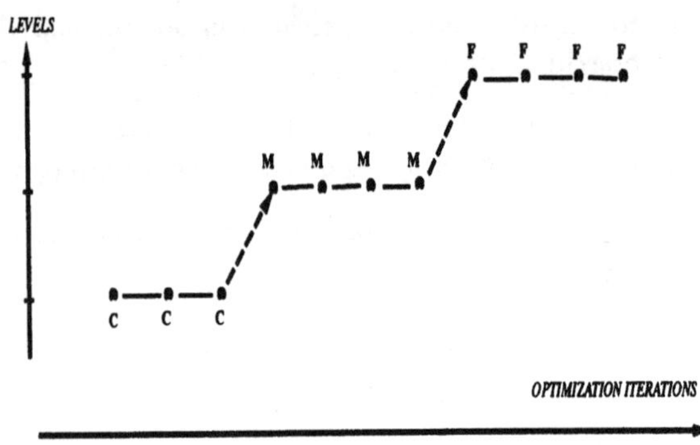

Figure 5 : Sketch of nested iteration strategy on 3 levels (C for coarse, M for medium, F for fine)

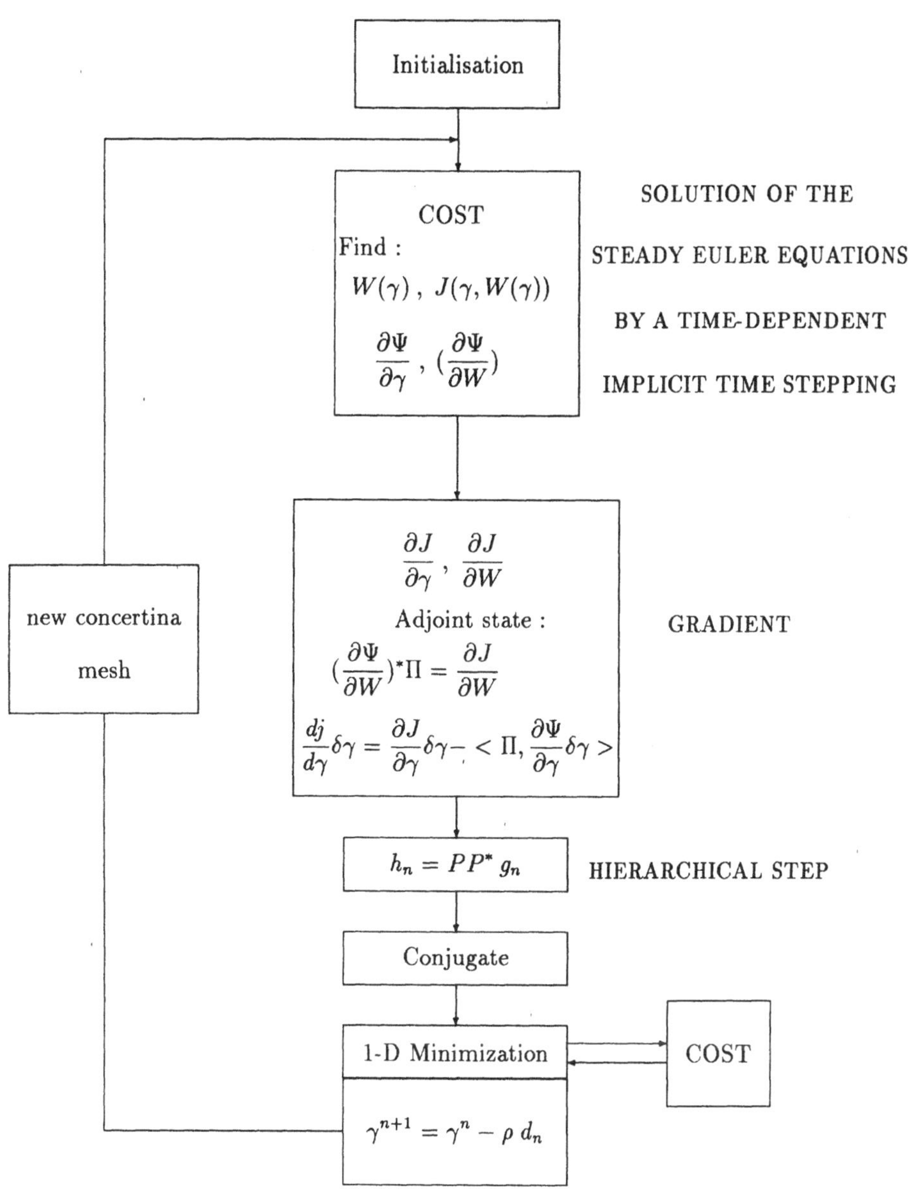

Figure 6 : Chart of optimization gradient algorithm

RESULTS

BASIC FEATURES OF THE EXPERIMENT

The chosen mesh has 423 nodes with 31 nodes on the controlled boundary; except the deformation related to the shape, the mesh is regular (see Figure 2). A hierarchical method associated with a nested cubic interpolation is applied. The calculations have been performed on a CONVEX C210.
We present here two types of numerical experiments. First we consider an inverse problem, i.e. we try to minimize a cost functional j defined as follows :

$$j(\gamma) = \int_0^2 (P_\gamma - P_{desired})^2 \, dx$$

in which $P_{desired}$ is a target pressure distribution obtained from the results of a direct computation performed for the design given shape where iterative residual is driven to machine zero. This problem is applied in the two first experiments.
The second considered problem is an optimization problem where the used cost functional is $j = \int_0^2 |\frac{\partial P}{\partial x}|^\alpha |_{y=0} \, dx$ where α is equal to 4.

Test Case T_2 and Test Case T_3 correspond respectively to a subsonic flow and a transonic flow. We refer to the specifications of the workshop for more precision on Test Cases T_1 , T_2 and T_3.

TEST CASE T1

We prescribe, here, a pressure distribution on the fixed symmetry axis (at the bottom of the computational domain: $y = 0$). The applied hierarchical method is a nested iteration strategy on five levels (1, 3, 7, 15 and 31 parameters) with a conjugation step. The cost functional converges rather fastly to zero, as sketched in Figures 7 and 8. Moreover, since the first iteration, we obtain a shape that is very close to the desired shape, as it is shown in Figure 12. We present in Figures 7 to 10 the convergence results and the isomach contours for the obtained final shape are plotted in Figure 11.
The global CPU time is 15320 seconds.

TRANSONIC INVERSE PROBLEM

We consider now a transonic flow in the same nozzle geometry as in test case T_1 (the farfield Mach number is equal to 0.5 instead of 0.2); but we take, here, the pressure distribution P_γ on the upper variable wall ($y = \gamma(x)$). The same hierarchical strategy is applied.
The general features of the obtained flow (isomach contours) are shown in Figure 13.
In Figure 14, we plot the following quantities: (1) the logarithm of the cost, (2) the

logarithm of l_2 gradient norm and (3) the logarithm of the deviation from the optimal shape ($||\gamma - \gamma_{desired}||_2$) as functions of the number of cost evaluations. These values are normalized by the initial value like in Test Case T_1.
We see, here, that the transonic regime doesn't spoil the results. We can notice a convergence improvement, but it is principally due to the choice of pressure distribution. P_γ is prescribed not on the symmetrical axis but on the wall: this choice is more natural in order to optimize the wall shape. The global CPU time (to decrease ten orders of magnitude the cost function) is 3850 seconds.

TEST CASE T2

We present in Figures 15 to 16 the convergence results. This problem appears rather stiff and the use of a large number of shape parameters was observed to lead to oscillations in the final shape.
In this subsonic case with an initial symmetrical geometry, the final shape should be symmetrical. Thus, we impose the symmetry and we use only few shape parameters in order to prevent oscillations formation. In fact, a Full V-cycle strategy without conjugaison is employed with only two levels (2 and 6 parameters).
The global CPU time is 1700 seconds.

TEST CASE T3

We present in Figures 17 to 18 the convergence results.
The extra difficulty provided by the transonic character of the flow has not a great influence on convergence; however, the shock induces strong pressure gradient. We observe an inflation of the shape in the region of the shock that tends to compensate it. Beginning with a transonic flow and a symmetric shape, we finally obtain a subsonic flow and a strongly dissymmetrical shape.
The global CPU time is 4000 seconds.

The obtained final shapes (with Mach contours lines) for Test Cases T_2 and T_3 are plotted respectively in Figures 19 and 20.

CONCLUSIONS

The information brought by the exact gradient for a small extra CPU effort is of interest as far as we can compute this exact gradient; this is proved to be possible with a first-order accurate upwind approximation; for further extension to second-order and Navier-Stokes, we think that automated symbolic differentiation will permit to do it with reasonable programming efforts.

The hierarchical method brings also an increment in efficiency which can be combined with different methods of optimization; it relies on a set of mathematical parametrizations that seem to be as good as those derive from smooth continuous curves. It can be also efficiently combined with approximate-gradient approaches.

The availability of both optimality conditions and a hierarchy of parametrization will permit to extend these methods to faster methods relying on simultaneous and hierarchical solution (cf [3]) of the global optimality system (state equations and gradient vanishing).

ACKNOWLEDGEMENTS

This work was supported by Brite Euram 1082 Contract (Optimum Design in Aerodynamics). I whish to thank my colleagues Loula Fézoui and Hervé Stève for yielding their implicit Euler finite-element code and I also express my thanks to Alain Dervieux for helpful advices.

REFERENCES

[1] F. BEUX, A. DERVIEUX " *Exact-gradient shape optimization of a 2-D Euler flow* ", Finite Element in Analysis and Design (Eds Elsevier), vol. 12, p. 281-302 (1992).

[2] F. BEUX, A. DERVIEUX *"A hierarchical approach for shape optimization"*, INRIA Research Report N° 1868 (1993), submitted to Engineering Computations.

[3] F. BEUX, N. MARCO, A. DERVIEUX *"Shape optimization for complex flows: toward coupled approaches"*, INRIA Contribution to BE-1082 (Optimum Design in Aerodynamics) 24th-month synthesis, subtasks 1.5, 2.5 (1992).

[4] A. BRANDT, " *Multigrid techniques* ", Guide with Applications to Fluid Dynamics, G.M.D-Studien, N° 85 (1984).

[5] W.L. BRIGGS, *A multigrid tutorial*, SIAM, Philadelphia (1987).

[6] H. STEVE, L. FEZOUI *"Décomposition d'un flux de Van-Leer pour résoudre les équations d'Euler par un schéma décentré en maillage non structuré", INRIA Research Report N° 825 (1988).*

[7] B. VAN LEER *"Flux Vector Splitting for the Euler Equations"*, Lecture notes in Physics, vol. 170, p. 405-512 (1982).

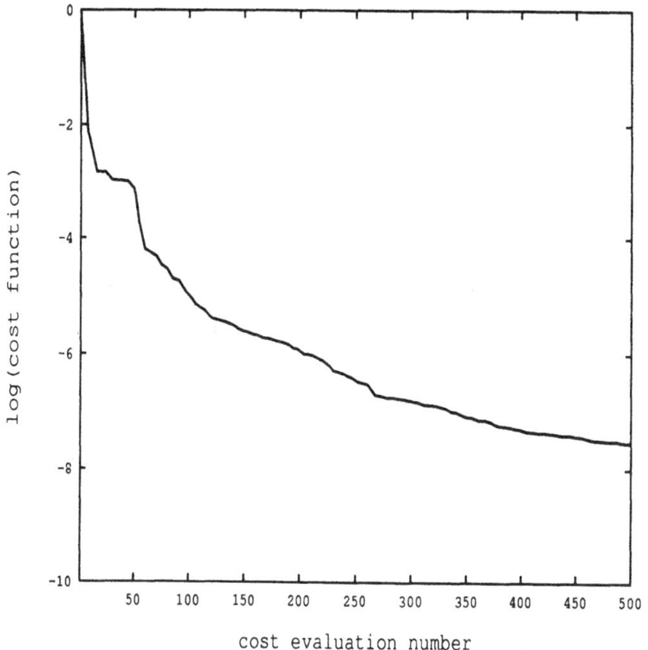

Figure 7 : T1: log of the cost as a function of cost evaluation number

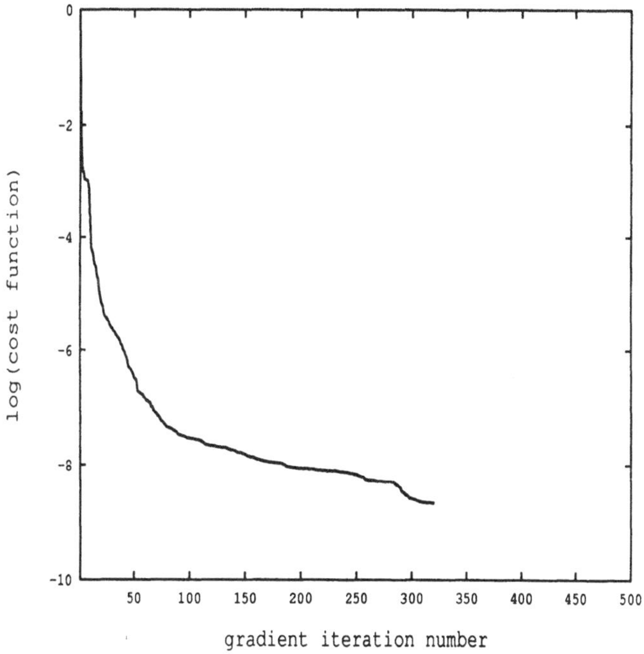

Figure 8 : T1: log of the cost as a function of gradient iteration number

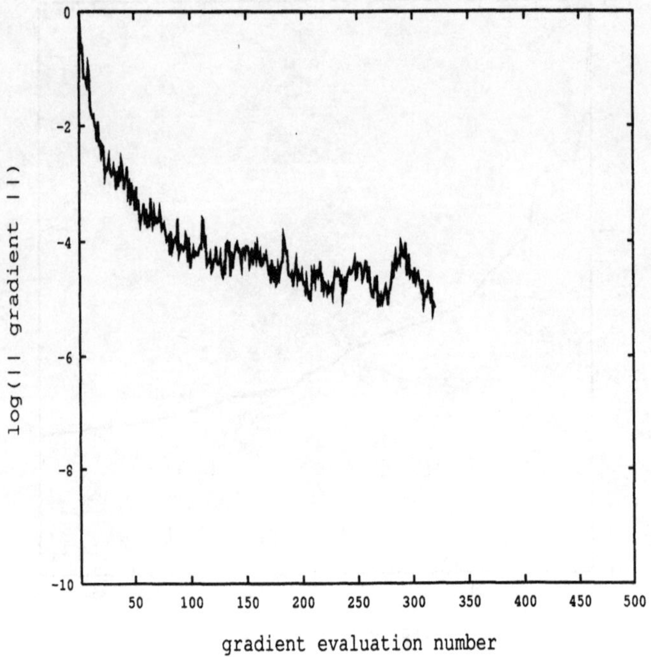

Figure 9 : T1: log of the gradient norm as a function of gradient evaluation number

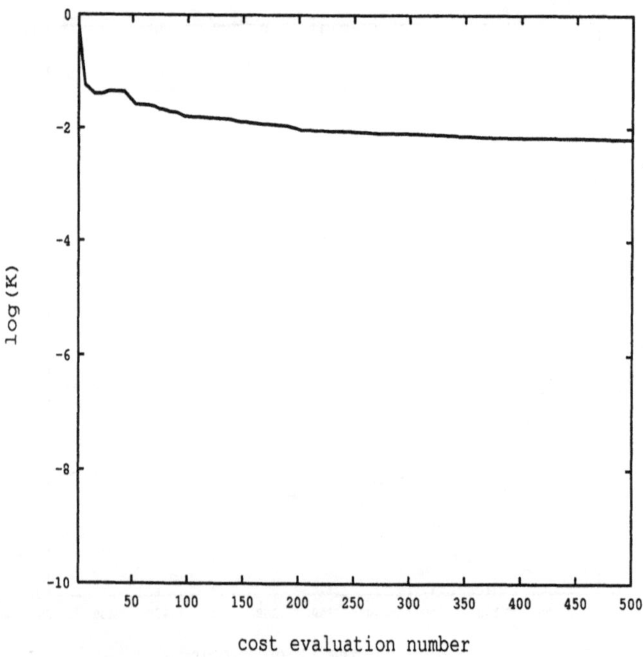

Figure 10 : T1: log of the deviation with optimal shape as a function of cost evaluation number ($K = ||\gamma - \gamma_{desired}||_2$)

30 ISOMACH
Isovalues spacing : 0.76959E-02
MIN = 0.17469 , MAX = 0.39787

Figure 11 : T1: Distribution of the final nozzle shape

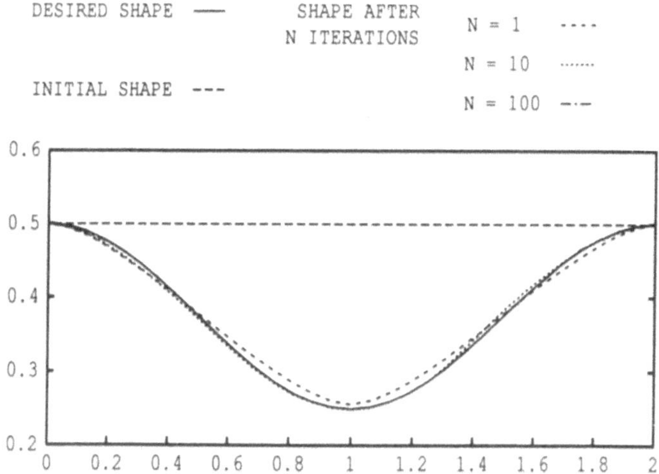

Figure 12 : T1: Successives shapes

Figure 13 : Transonic inverse problem: distribution of the final nozzle shape

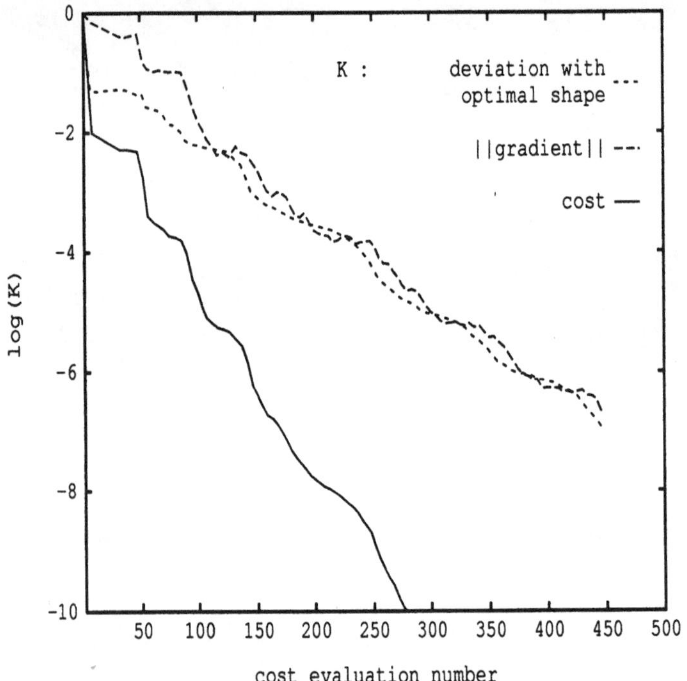

Figure 14 : Transonic inverse problem: convergence history

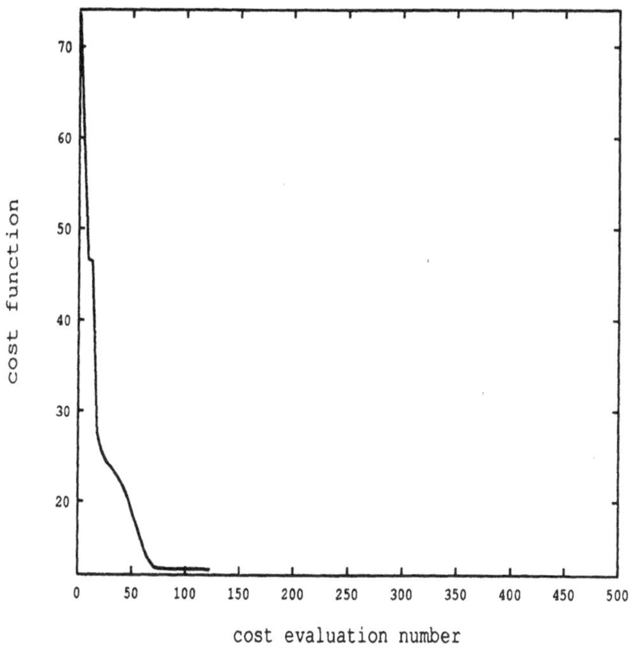

Figure 15 : T2: Cost functional as a function of cost evaluation number

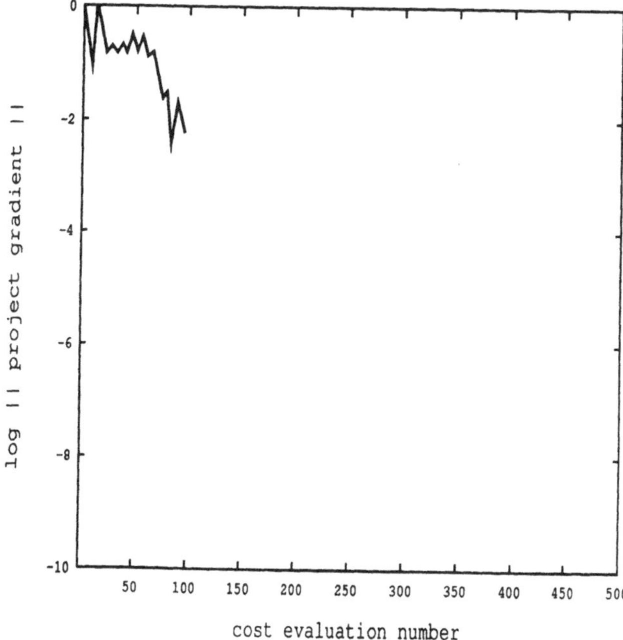

Figure 16 : T2: log of the projected gradient norm as a function of cost evaluation number

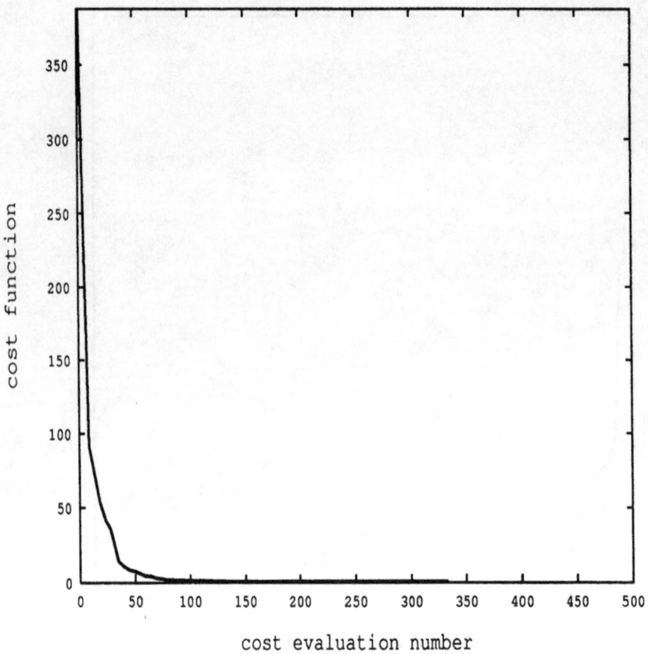

Figure 17 : T3: Cost functional as a function of cost evaluation number

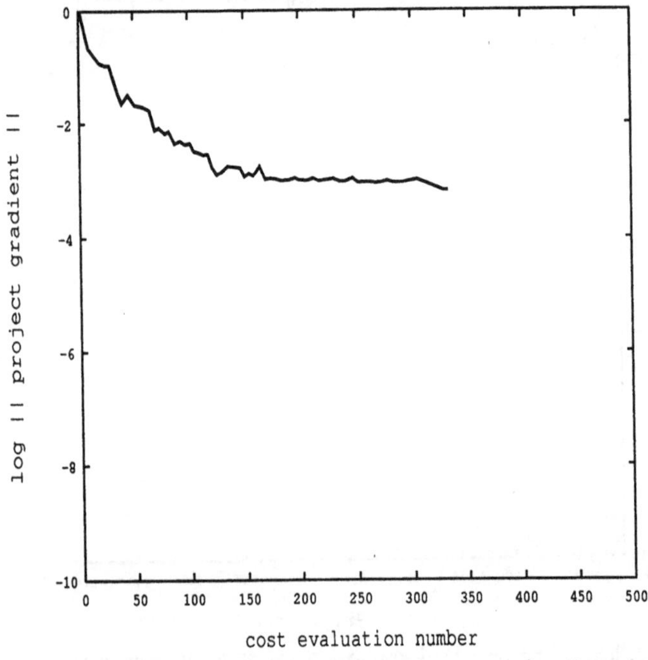

Figure 18 : T3: log of the projected gradient norm as a function of cost evaluation number

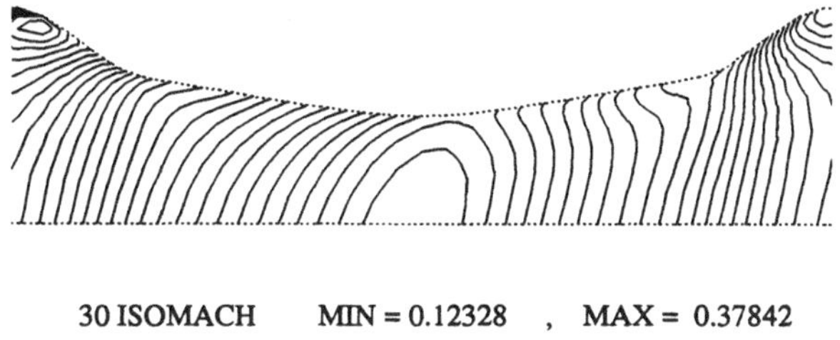

30 ISOMACH MIN = 0.12328 , MAX = 0.37842

Figure 19 : T2: Distribution of the obtained nozzle shape

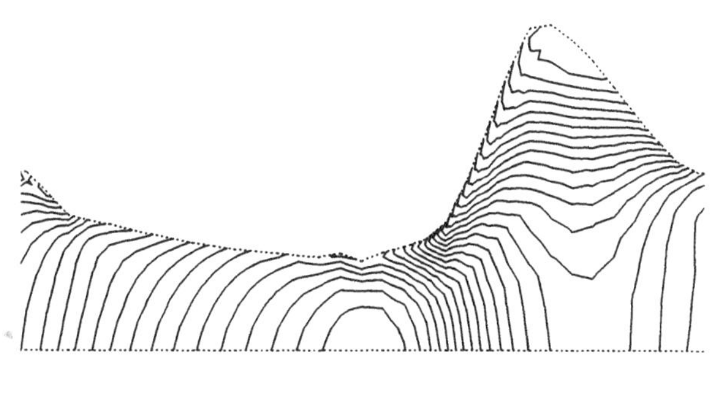

30 ISOMACH MIN = 0.26477E-02 , MAX = 0.63504

Figure 20 : T3: Distribution of the obtained nozzle shape

AERODYNAMIC SHAPE OPTIMIZATION USING AUTOMATIC ADAPTIVE REMESHING

G. BUGEDA, E. OÑATE, D. JOANNAS

Escola Tècnica Superior d'Enginyers de Camins, Canals i Ports
Universitat Politècnica de Catalunya
Módulo C1; Campus Norte UPC; Gran Capitán s/n;08034 Barcelona;Spain

SUMMARY

This work presents a new methodology based on the use of adaptive mesh refinement (AMR) techniques in the context of shape optimization problems analyzed by the Finite Element Method. A suitable and very general technique for the parametrization of the optimization problem using B-splines to define the boundary is first presented. Then, mesh generation using the advancing front method, the estimation of error and the mesh refinement criterion are studied in the context of shape optimization problems. In particular, the sensitivities of the different ingredients ruling the problem (B-splines, finite element mesh, flow behaviour, and error estimator) are studied in detail. The sensitivities of the finite element mesh and the error estimator allow their projection from one design to the next, thus leading to an "a priori knowledge" of the error distribution on the new design without the need of any additional analysis. This information allows to build up a finite element mesh for the new design with an specified and controlled level of error. The robustness and reliability of the proposed methodology is checked out with some 2D application examples.

1. INTRODUCTION

From a mathematical point of view the treatment of an optimization or an inverse problem can be viewed as the minimization of a real function $f(\boldsymbol{x})$ depending on a set of variables ($\boldsymbol{x} = \{x_i\}\ ; i = 1, ..., n$). Each set defines a different design and the problem consists of finding those $\boldsymbol{x}$ values defining the best design.

The algorithms for the solution of the minimization problem are, typically, iterative and they involve the computation of the derivatives (sensitivities) of the objective function with respect to the design variables. Besides, in each step of the process, the computation of the f values and their sensitivities is needed. In many cases, as those considered in this work, the computations are performed via a finite element analysis which provides the aerodynamic response of each design, and a way to compute the corresponding sensitivities. The definition of each design in terms of the $\boldsymbol{x}$ variables is called the "parametrization" of the optimum design problem.

The author's contribution to the EUROPT I [4,5] project has been the development of a new methodology for optimization and inverse aerodynamic problems including an adaptive remeshing procedure. The main characteristics of this methodology are:

- The control of the error involved in the numerical solution of the flow equations for each design using an error estimator. This is used for the generation of a new mesh for each design.

- The use of first and second order *exact* sensitivity analysis of all the magnitudes involved in the process, including the coordinates of the mesh, the velocities, the objective function and the error estimator.
- The *projection* of the information obtained with one design to the next one, using the mentioned sensitivity analysis. This projection is used to define the characteristics of the mesh which will ensure good quality results for the next design.
- The parametrization of the problem using B-splines. The coordinates of the points used to define a B-spline are used as design variables.

The proposed methodology has been implemented in a computer code using an incompressible potential flow model. The code has been tested by solving different 2D applications like the workshop test cases defined for the EUROPT I project [19].

2. THE PROPOSED METHODOLOGY

A general scheme of the proposed methodology is shown in Figure 1. It consists of a series of modules each one corresponding to a specific task. In the next sections some of these modules are discussed in detail.

For each design it is necessary to compute the sensitivities of the objective function. The sensitivity analysis for the whole problem is performed step by step following the same path as the flow analysis. This path indicates the dependence of each quantity used in the analysis with respect to the rest of the quantities previously employed. For example, the expression of the finite element matrix depends on the nodal coordinates, so that, following the chain rule for derivatives, the finite element matrix sensitivities can be expressed in terms of the nodal coordinates sensitivities. Thus, it is necessary to compute these sensitivities (mesh sensitivities) prior to that of the finite element matrix sensitivities.

First-order and second-order sensitivity analyses have been used in the implementation of the proposed methodology. The sensitivity analysis provides directional derivatives of any quantity. In the next paragraphs the $\boldsymbol{s}$ will denote a unit vector in the design variables space ($\boldsymbol{x} = \{x_1, x_2, ..., x_i, ..., x_n\}$), and derivatives will be computed in the $\boldsymbol{s}$ direction. For instance, to obtain the sensitivities with respect to a specific design variable x_i, $\boldsymbol{s}$ has to be the unit vector corresponding to the x_i direction (i.e. $\boldsymbol{s} = \{0, 0, ..., 1, ..., 0\}$).

The sensitivities of any quantity will be used to project its value from one design to the next one when the design variables are modified. For example, let us assume that $f(\boldsymbol{x}^k)$ is the objective function value at the k-th iteration of the optimization process. If the design variables are modified in the form $\boldsymbol{x}^{k+1} = \boldsymbol{x}^k + \theta^k \boldsymbol{s}^k$, the value of the objective function can be projected to the next design by means of a standard Taylor expansion:

$$f(\boldsymbol{x}^{k+1}) = f(\boldsymbol{x}^k + \theta \boldsymbol{s}^k) \approx f(\boldsymbol{x}^k) + \theta^k \frac{\partial f}{\partial \boldsymbol{s}} + \frac{1}{2} \theta^{k2} \frac{\partial^2 f}{\partial \boldsymbol{s}^2} \tag{1}$$

The same applies to any other magnitude to be projected.

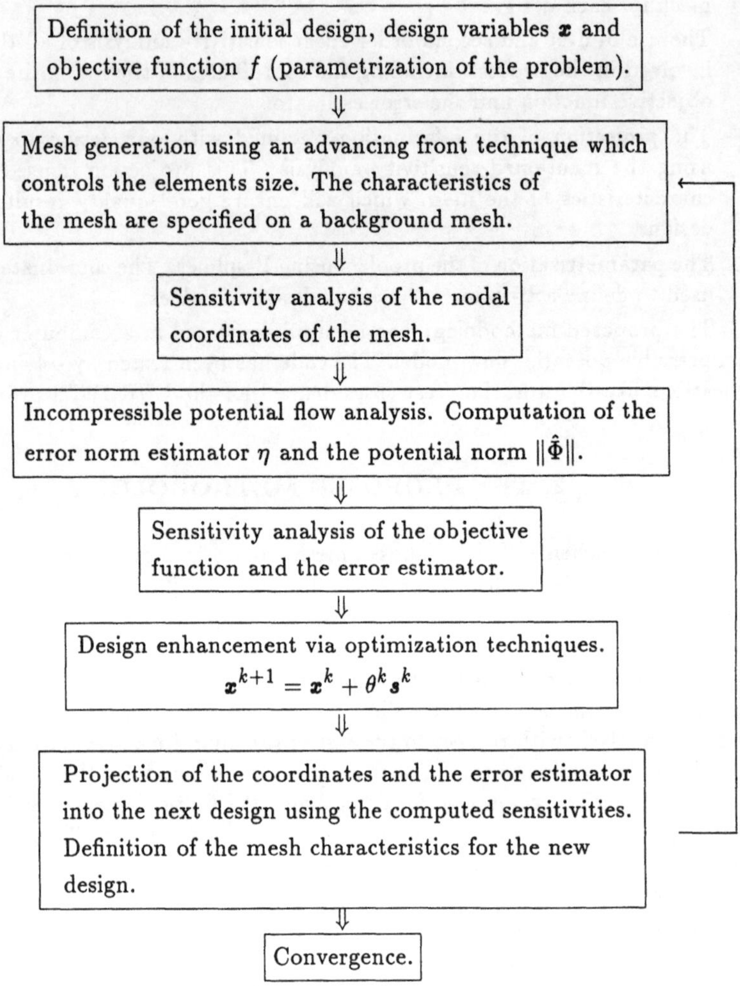

Definition of the initial design, design variables $\boldsymbol{x}$ and objective function f (parametrization of the problem).

⇓

Mesh generation using an advancing front technique which controls the elements size. The characteristics of the mesh are specified on a background mesh.

⇓

Sensitivity analysis of the nodal coordinates of the mesh.

⇓

Incompressible potential flow analysis. Computation of the error norm estimator η and the potential norm $\|\hat{\Phi}\|$.

⇓

Sensitivity analysis of the objective function and the error estimator.

⇓

Design enhancement via optimization techniques.
$\boldsymbol{x}^{k+1} = \boldsymbol{x}^k + \theta^k \boldsymbol{s}^k$

⇓

Projection of the coordinates and the error estimator into the next design using the computed sensitivities. Definition of the mesh characteristics for the new design.

⇓

Convergence.

Figure 1. General scheme of the proposed methodology.

3. PARAMETRIZATION OF THE PROBLEM

The description of each design geometry is done by using "definition points" which specify some interpolation curves. The curves used here are parametric B-splines. The general expression of a closed B-spline for q points is [1,6]:

$$\boldsymbol{r}(t) = \sum_{l=0}^{q} \boldsymbol{r}_l N_{4,l+1}(t) \tag{2}$$

where $\boldsymbol{r}(t)$ is the position vector depending on a parametric variable t. The coordinates of the definition points are recovered using $t = 0, 1, 2, ...$ The curve is expressed as a linear combination of $q+1$ normalized fourth order (cubic) B-splines [1,6]. The $\boldsymbol{r}_l$

coefficients are the coordinates of the so called polygon definition points [1,6] and they are found by using the coordinates of the definition points. The degree of continuity of a cubic B-spline is C^2. By using eq. (2) the coordinates of the definition points and some additional conditions about slopes and curvatures the following equations system can be found:

$$\boldsymbol{V} = \boldsymbol{N}\boldsymbol{R} \tag{3}$$

where $\boldsymbol{V}$ is a vector containing the imposed conditions at the definition points, $\boldsymbol{N}$ is a matrix containing some terms corresponding to the values of the polynomials that define each B-spline, and the $\boldsymbol{R}$ vector contains the coefficients $\boldsymbol{r}_i$ to be computed. Details of this process can be found in [1,6].

The first and second order sensitivities of $\boldsymbol{R}$ along a direction $\boldsymbol{s}$ in the design variable space are given by:

$$\frac{\partial \boldsymbol{R}}{\partial \boldsymbol{s}} = \boldsymbol{N}^{-1}\left(\frac{\partial \boldsymbol{V}}{\partial \boldsymbol{s}} - \frac{\partial \boldsymbol{N}}{\partial \boldsymbol{s}}\boldsymbol{R}\right) \quad , \quad \frac{\partial^2 \boldsymbol{R}}{\partial \boldsymbol{s}^2} = \boldsymbol{N}^{-1}\left(\frac{\partial^2 \boldsymbol{V}}{\partial \boldsymbol{s}^2} - \frac{\partial^2 \boldsymbol{N}}{\partial \boldsymbol{s}^2}\boldsymbol{R} - 2\frac{\partial \boldsymbol{N}}{\partial \boldsymbol{s}}\frac{\partial \boldsymbol{R}}{\partial \boldsymbol{s}}\right). \tag{4}$$

The derivatives of $\boldsymbol{V}$ with respect the coordinates of the definition points chosen as design variables can be very easily computed. Vectors $\frac{\partial \boldsymbol{R}}{\partial \boldsymbol{s}}$ and $\frac{\partial^2 \boldsymbol{R}}{\partial \boldsymbol{s}^2}$ will contain the terms $\frac{\partial \boldsymbol{r}_i}{\partial \boldsymbol{s}}$ and $\frac{\partial^2 \boldsymbol{r}_i}{\partial \boldsymbol{s}^2}$, respectively [1].

Finally, the sensitivities of the coordinates of any point on the interpolation curve corresponding to $t = constant$ are obtained by:

$$\frac{\partial \boldsymbol{r}(t)}{\partial \boldsymbol{s}} = \sum_{i=0}^{q} \frac{\partial \boldsymbol{r}_i}{\partial \boldsymbol{s}} N_{4,i+1}(t) \quad , \quad \frac{\partial^2 \boldsymbol{r}(t)}{\partial \boldsymbol{s}^2} = \sum_{i=0}^{q} \frac{\partial^2 \boldsymbol{r}_i}{\partial \boldsymbol{s}^2} N_{4,i+1}(t). \tag{5}$$

4. MESH GENERATION AND SENSITIVITY ANALYSIS

The chosen algorithm to generate the mesh is the well known advancing front method [12,13]. This technique is ideal to generate non structured triangular meshes.

The characteristics of the desired mesh are specified via a background mesh over which nodal values of the size parameters δ are defined and interpolated using the shape functions. For the first design the background mesh has to be defined by hand. For subsequent designs the background mesh will coincide with the mesh projected into this design from the previous one. This projection will be described later.

Once the sensitivities of the coordinates of each boundary node are known, it is also possible to compute the sensitivities of the coordinates of each internal nodal point (mesh sensitivities). These sensitivities are used to asses how the mesh moves when the design variables change.

For non-structured meshes the sensitivity analysis can be obtained by studying the variations of the mesh due to changes in the design variables. There are many different ways to define the movement of the mesh in terms of the design variables. It is possible to consider a simple analogous elastic medium defining the mesh movement. This is the case of the "spring analogy" where each element side is regarded as a spring connecting two nodes. The force produced by each spring is proportional to its length. The solution of the equilibrium problem in the spring analogy is simple but expensive and it involves to solve a linear system of equations with two of degrees of freedom for each node.

The spring analogy problem can be solved iteratively using a Laplacian smoothing, as it has been done in the present work. This technique is frequently used to improve the quality of non-structured meshes. It consists on the iterative modification of the nodal coordinates of each interior node by placing it at the center of gravity of the adjacent nodes. For each iteration the expression of the new position vector of each node $\boldsymbol{r}_i$ is given by:

$$\boldsymbol{r}_i = \frac{\sum_{j=1}^{m_i} \boldsymbol{r}_j}{m_i} \tag{6}$$

where $\boldsymbol{r}_j$ are the position vectors of the m_i nodes connected with the i-th node.

The solution of the spring analogy problem with a prescribed error tolerance requires to check the solution after each smoothing cycle. Taking into account that the described iterative process is only a way to obtain mesh sensitivities, rather than the solution of the equilibrium problem itself, rigorous convergence conditions are not needed. For this reason the number of smoothing cycles to be applied can be fixed a priori. In the examples presented below we have checked that 50 iterations are enough to ensure a good quality of the results.

The first-order and higher mesh sensitivity analysis in any direction of the design variables space, $\boldsymbol{s}$, are obtained by differentiating eq. (6) with respect to $\boldsymbol{s}$ for each cycle, i.e.

$$\frac{\partial \boldsymbol{r}_i}{\partial \boldsymbol{s}} = \frac{\sum_{j=1}^{m_i} \frac{\partial \boldsymbol{r}_j}{\partial \boldsymbol{s}}}{m_i} \quad , \qquad \frac{\partial^2 \boldsymbol{r}_i}{\partial \boldsymbol{s}^2} = \frac{\sum_{j}^{m_i} \frac{\partial^2 \boldsymbol{r}_j}{\partial \boldsymbol{s}^2}}{m_i} \ . \tag{7}$$

5. FLOW ANALYSIS AND ERROR ESTIMATION

The basic equations for the analysis of the flow around a profile using an incompressible potential model with lifting involve a "continuous" potential Φ_0 and a "non-continuous" potential Φ_1 which are combined in order to accomplish the Kutta-Joukowski condition [14].

Let us consider a domain Ω where the flow problem is defined. A cut Σ between the trailing edge Te and the boundary of Ω is defined (see Figure 2). The upper and lower sides of the cut are noted Σ^+ and Σ^-, respectively.

The corresponding equations for the "continuous" and the "non continuous" potentials are [14]:

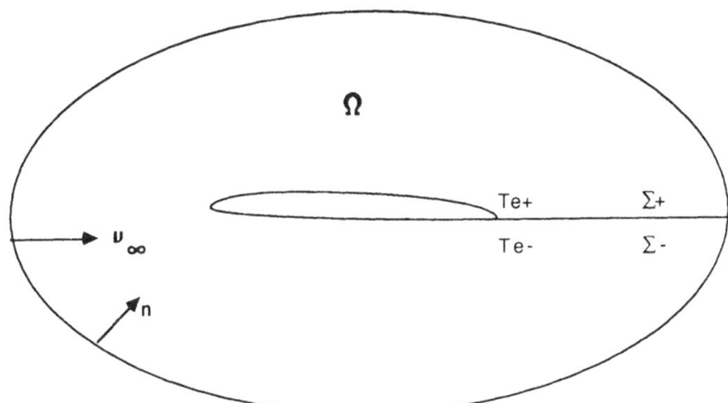

Figure 2. Analysis domain for the potential flow with lifting problem.

$$\begin{cases} \Delta\Phi_0 = 0 \\ \Phi_0|_{\Sigma^+} - \Phi_0|_{\Sigma^-} = 0 \\ \frac{\partial\Phi_0}{\partial n^+}|_{\Sigma^+} + \frac{\partial\Phi_0}{\partial n^-}|_{\Sigma^-} = 0 \\ \frac{\partial\Phi_0}{\partial n}|_{\partial\Omega} = \boldsymbol{v}_\infty \boldsymbol{n} \\ \Phi_0(Te^-) = 0 \end{cases} \qquad \begin{cases} \Delta\Phi_1 = 0 \\ \Phi_1|_{\Sigma^+} - \Phi_1|_{\Sigma^-} = 1 \\ \frac{\partial\Phi_1}{\partial n^+}|_{\Sigma^+} + \frac{\partial\Phi_1}{\partial n^-}|_{\Sigma^-} = 0 \\ \frac{\partial\Phi_1}{\partial n}|_{\partial\Omega} = 0 \\ \Phi_0(Te^-) = 0\,. \end{cases} \tag{8}$$

Both potentials are combined to get the final solution $\Phi = \Phi_0 + \lambda\Phi_1$ where λ is obtained from the Kutta-Joukowski condition [14] which ensures the continuity of the upper and lower velocities at the trailing edge.

The velocities at each point are obtained from the gradient of the potentials $\boldsymbol{v} = \nabla\Phi = \nabla\Phi_0 + \lambda\nabla\Phi_1$.

If more than one airfoil is involved a non continuous potential is defined for each geometry and the Kutta-Joukowski condition is applied to each trailing edge.

The discretization of eqs. (8) for each potential Φ using a finite element approximation, i.e. $\Phi \approx \hat{\Phi} = \sum_i N_i a_i = \boldsymbol{Na}$, leads to a linear system of equations, i.e.

$$\boldsymbol{Ka} = \boldsymbol{f} \quad with \quad \begin{cases} \boldsymbol{K} = \sum_e \boldsymbol{K}_e \\ \boldsymbol{K}_e = \int_{\Omega_e} \boldsymbol{B}^T \boldsymbol{B} d\Omega \\ \boldsymbol{f} = \int_{\Gamma_e} \boldsymbol{N}\boldsymbol{v}^T \boldsymbol{n} d\Gamma. \end{cases} \tag{9}$$

Matrix $\boldsymbol{B}$ contains the cartesian derivatives of the shape functions N_i and it can be used to obtain the velocities corresponding to the approximated potential $\hat{\boldsymbol{v}} = \boldsymbol{B}\hat{\boldsymbol{\Phi}}$.

For incompressible potential problems the "energy" of the exact solution can be defined as:

$$\|U\| = [\int_\Omega \boldsymbol{v}^T \boldsymbol{v} d\Omega]^{1/2}. \tag{10}$$

The error of this "energy" can be estimated using the popular error estimator developed by Zienkiewicz and Zhu [15–18] for structural problems. The extension to potential flow problems requires the definition of the following error norm [3,4]:

$$\|e\| = [\int_{\Omega} [\boldsymbol{v} - \hat{\boldsymbol{v}}]^T [\boldsymbol{v} - \hat{\boldsymbol{v}}] d\Omega]^{1/2} \tag{11}$$

where $\boldsymbol{v}$ are the "exact" velocities, $\hat{\boldsymbol{v}}$ are the velocities obtained from the finite element solution and Ω is the flow analysis domain.

Since the exact velocities are usually not known they are approximated by $\boldsymbol{v} \simeq \boldsymbol{v}^* = \boldsymbol{N}\bar{\boldsymbol{v}}^*$ where $\bar{\boldsymbol{v}}^*$ are nodal values obtained by simple nodal averaging of the finite element values, local or global least squares smoothing, or other adequate projection methods [15–18]. A simple approach is to use a global nodal smoothing with a "mass" matrix giving the nodal smoothed values, $\bar{\boldsymbol{v}}^*$, as

$$\bar{\boldsymbol{v}}^* = \boldsymbol{M}^{-1} \int_{\Omega} \boldsymbol{N}_v \hat{\boldsymbol{v}} d\Omega \tag{12}$$

where $\boldsymbol{N}_v$ are the chosen velocity interpolating functions giving a smooth nodal velocity field [15–18] and $M_{ij} = \int_{\Omega} N_{v_i} N_{v_j} d\Omega$. Eq. (12) can be obviously applied to solve independently for each individual velocity component.

The "energy" of the exact solution is estimated as

$$\|U\| \approx [\int_{\Omega} \boldsymbol{v}^{*T} \boldsymbol{v}^* d\Omega + \int_{\Omega} [\boldsymbol{v}^* - \hat{\boldsymbol{v}}]^T [\boldsymbol{v}^* - \hat{\boldsymbol{v}}] d\Omega]^{1/2} \tag{13}$$

Both $\|e\|^2$ and $\|U\|^2$ can be evaluated as sum of their respective element contributions.

6. SENSITIVITY ANALYSIS OF THE OBJECTIVE FUNCTION AND THE ERROR ESTIMATOR

The objective function will be normally expressed in terms of the velocities, pressures, etc. obtained from the flow model. To compute the sensitivity of the objective function it is necessary to evaluate the sensitivity of all the magnitudes involved in the flow analysis.

The exact sensitivity analysis of all the integral expressions involved in the finite element discretization of the incompressible potential flow model can be obtained by direct derivation of eqs. (9). This provides the sensitivities of all magnitudes in terms of the mesh sensitivities previously obtained. Details of this process are described in references [1,7,8]. The corresponding expressions can be separately applied to each potential (Φ_0, Φ_1,...).

For a single airfoil, the sensitivity analysis of the total potential $\hat{\Phi} = \hat{\Phi}_0 + \lambda\hat{\Phi}_1$ involves a combination of the sensitivities of Φ_0, Φ_1 and λ.

$$\frac{\partial \hat{\Phi}}{\partial \boldsymbol{s}} = \frac{\partial \hat{\Phi}_1}{\partial \boldsymbol{s}} + \lambda \frac{\partial \hat{\Phi}_2}{\partial \boldsymbol{s}} + \frac{\partial \lambda}{\partial \boldsymbol{s}} \hat{\Phi}_2 \quad , \quad \frac{\partial^2 \hat{\Phi}}{\partial \boldsymbol{s}^2} = \frac{\partial^2 \hat{\Phi}_1}{\partial \boldsymbol{s}^2} + \lambda \frac{\partial^2 \hat{\Phi}_2}{\partial \boldsymbol{s}^2} + 2 \frac{\partial \lambda}{\partial \boldsymbol{s}} \frac{\partial^2 \hat{\Phi}_2}{\partial \boldsymbol{s}^2} + \frac{\partial^2 \lambda}{\partial \boldsymbol{s}^2} \hat{\Phi}_2 . \tag{14}$$

The terms $\partial\lambda/\partial\boldsymbol{s}$ and $\partial^2\lambda/\partial\boldsymbol{s}^2$ can be obtained by direct derivation of the Kutta-Joukowski condition.

The same operation can be performed to compute the sensitivities of the velocities at the integration points ($\hat{\boldsymbol{v}}$) and the nodal points ($\bar{\boldsymbol{v}}^*$) i.e.

$$\frac{\partial \hat{\boldsymbol{v}}}{\partial \boldsymbol{s}} = \frac{\partial \hat{\boldsymbol{v}}_1}{\partial \boldsymbol{s}} + \lambda \frac{\partial \hat{\boldsymbol{v}}_2}{\partial \boldsymbol{s}} + \frac{\partial \lambda}{\partial \boldsymbol{s}} \hat{\boldsymbol{v}}_2 \quad , \quad \frac{\partial^2 \hat{\boldsymbol{v}}}{\partial \boldsymbol{s}^2} = \frac{\partial^2 \hat{\boldsymbol{v}}_1}{\partial \boldsymbol{s}^2} + \lambda \frac{\partial^2 \hat{\boldsymbol{v}}_2}{\partial \boldsymbol{s}^2} + 2 \frac{\partial \lambda}{\partial \boldsymbol{s}} \frac{\partial^2 \hat{\boldsymbol{v}}_2}{\partial \boldsymbol{s}^2} + \frac{\partial^2 \lambda}{\partial \boldsymbol{s}^2} \hat{\boldsymbol{v}}_2 \tag{15}$$

$$\frac{\partial \bar{\boldsymbol{v}}^*}{\partial \boldsymbol{s}} = \frac{\partial \bar{\boldsymbol{v}}^*{}_1}{\partial \boldsymbol{s}} + \lambda \frac{\partial \bar{\boldsymbol{v}}^*{}_2}{\partial \boldsymbol{s}} + \frac{\partial \lambda}{\partial \boldsymbol{s}} \bar{\boldsymbol{v}}^*{}_2 \; , \; \frac{\partial^2 \bar{\boldsymbol{v}}^*}{\partial \boldsymbol{s}^2} = \frac{\partial^2 \bar{\boldsymbol{v}}^*{}_1}{\partial \boldsymbol{s}^2} + \lambda \frac{\partial^2 \bar{\boldsymbol{v}}^*{}_2}{\partial \boldsymbol{s}^2} + 2 \frac{\partial \lambda}{\partial \boldsymbol{s}} \frac{\partial^2 \bar{\boldsymbol{v}}^*{}_2}{\partial \boldsymbol{s}^2} + \frac{\partial^2 \lambda}{\partial \boldsymbol{s}^2} \bar{\boldsymbol{v}}^*{}_2 . \tag{16}$$

Very similar equations to (14), (15) and (16) hold for cases with more than one airfoil. Here there are as many λ parameters as different components where the Kutta-Joukowski condition is imposed. These parameters are obtained by solving a non linear system of equations, each one corresponding to the Kutta-Joukowski condition applied over one airfoil. To obtain the sensitivities of the λ parameters it is necessary to derivate this system of equations with respect to the velocities at the trailing edge of each airfoil.

The computation of C_p and its sensitivities in terms of the velocities for the case of an incompressible flow is trivial using the expression:

$$C_p = 1 - \left(\frac{|\boldsymbol{v}|}{|\boldsymbol{v}_\infty|} \right)^2 . \tag{17}$$

The sensitivities of the objective function can be obtained by its direct derivation with respect to the design variables. These sensitivities will be expressed in terms of the sensitivities of the nodal velocities.

The sensitivities of the error estimator and the "energy" of the solution can be obtained in terms of the sensitivities of the velocities by appropriate derivation of the integral expressions (11) and (13) [1,7,8].

7. DESIGN IMPROVEMENT

The objective function sensitivities are used to get improved values of the design variables by means of a minimization method. Depending on the optimization algorithm it may be necessary to use second order sensitivities. The design variables corresponding to the improved design will usually be found as:

$$\boldsymbol{x}^{k+1} = \boldsymbol{x}^k + \theta \boldsymbol{s}^k \tag{18}$$

where θ is an advance parameter.

The direction of change $\boldsymbol{s}^k$ has been obtained here using a BFGS Quasi-Newton method which only requires first order sensitivities of the objective function. The value of θ is obtained by a directional second order sensitivity analysis in the $\boldsymbol{s}^k$ direction. The objective function f can be approximated along this direction using a second order Taylor expansion similar to eq. (1) which minimization provides the value of θ. Details of this algorithm can be found in [7].

8. PROJECTION TO THE NEXT DESIGN AND DEFINITION OF THE NEW MESH

Once the new design has been defined the new values of the error estimator, the "energy" and the coordinates of the mesh can be projected from the previous solution as:

$$(x,y)^{k+1} = (x,y)^k + \theta\Big(\frac{\partial x}{\partial s}, \frac{\partial y}{\partial s}\Big) + \frac{1}{2}\theta^2\Big(\frac{\partial^2 x}{\partial s^2}, \frac{\partial^2 y}{\partial s^2}\Big) \tag{19}$$

$$\|e\|^{2^{k+1}} = \|e\|^{2^k} + \theta\frac{\partial\|e\|^2}{\partial s} + \frac{1}{2}\theta^2\frac{\partial^2\|e\|^2}{\partial s^2} \tag{20}$$

$$\|U\|^{2^{k+1}} = \|U\|^{2^k} + \theta\frac{\partial\|U\|^2}{s} + \frac{1}{2}\theta^2\frac{\partial^2\|U\|^2}{\partial s^2}\,. \tag{21}$$

These projections provide a good approximation of each of above values for the next design previously to any new computation. In fact, the projected values provide the necessary information to perform a remeshing over the next design, even before any new computation is performed. In that sense, we have converted an error estimator computed "a posteriori" into an "a priori" error estimator.

This projection is very important because it allows the quality control of the meshes for each design without any new remeshing. Only one initial mesh is generated and this is used over each design. Thus, the extra cost involved in the control of the mesh quality is very cheap.

The projected values are used to create the background mesh information needed to generate the first mesh for the new design. This operation closes the iterative process which will lead to the 'enhanced" optimum design after convergence.

For the complete definition of the characteristics of a new mesh in the remeshing procedure it is necessary to use a mesh optimality criterion. In this work a mesh is considered as optimal when the error density is equally distributed across the volume, i.e. when $\frac{\|e\|_e^2}{\Omega_e} = \frac{\|e\|^2}{\Omega}$ is satisfied. The justification of this optimality criterion can be found in [9–11].

The combination of the optimality criterion and the error estimation allows to define the new element sizes. Previously, it is necessary to define the limit of the allowable global error percentage γ as:

$$\gamma = 100\frac{\|e\|}{\|\Phi\|} \approx 100\frac{\|e\|}{\sqrt{\|e\|^2 + \|\hat{\Phi}\|^2}}\,. \tag{22}$$

The desired error level for each element is:

$$\|e\|_e^d = \frac{\gamma}{100}\sqrt{(\|\hat{\Phi}\|^2 + \|e\|^2)\frac{\Omega_e}{\Omega}}\,. \tag{23}$$

The new element sizes $\bar{h}_e$ can be computed in terms of the old ones h_e using the expression:

$$h = \frac{h_e}{\xi_e^{1/p}} \tag{24}$$

where $\xi_e = \frac{\|e\|_e}{\|e\|_e^d}$ and p is the order of the shape function polynomials. For further details see [1–3,9–11].

9. APPLICATION EXAMPLES

The methodology developed in this work has been implemented in a computer code in order to check its reliability. Test cases T1, T2, T4 and T10 of the workshop on Optimum Design in Aerodynamics held in Barcelona on June 1992 [19] have been tackled and its corresponding results are presented below.

9.1 T1 workshop test case

This test case consists in recovering a nozzle starting from a straight line. The target shape is defined by the following sine curve:

$$y(x) = 0.375 + 0.125 sin(\pi(x - 1.500)). \tag{25}$$

The target pressure coefficient C_p^{target} has been obtained using the defined target shape and computing it with a finite element code using the incompressible potential model previously defined and adaptive remeshing. Quadratic triangular elements (P2) have been used for the analysis. The maximum error γ has been limited to 0.1%.

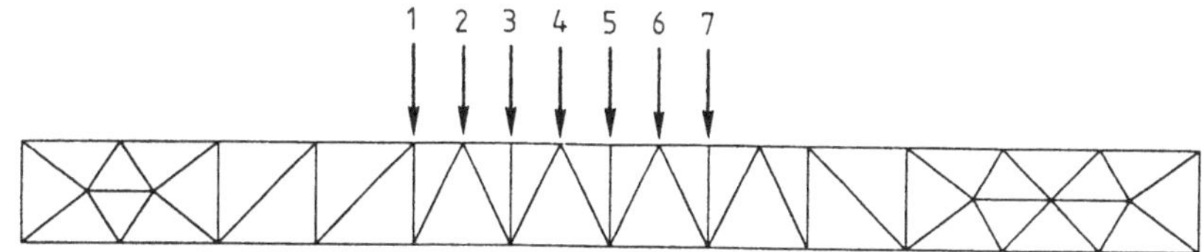

Figure 3. Initial shape, initial mesh and definition of the design variables.

The inverse problem has been solved using a minimization approach. The cost functional to be minimized has been defined as:

$$f = \int_0^2 (C_p(x) - C_p^{target})^2 dx\,. \tag{26}$$

Figure 4. Final shape and final mesh.

The geometry of each design has been defined using 7 design variables. This variables are the y coordinates of 7 points which are used to interpolate a B-spline along the top line. Figure 3 shows the initial shape, the position of the 7 design variables and the finite element mesh used for the analysis of initial design.

The iterative process has converged after 33 iterations. The final shape and the corresponding final mesh can be observed in Figure 4. Note the "quality" of the refinement automatically obtained in the zones where the solution error is greated. The whole problem has taken around 18 minutes of CPU running on a Personal Iris 35TG workstation.

The evolution of the normalized cost functional f during the process can be seen in Figure 5. Note that only one cost evaluation for each iteration is performed in the proposed methodology. Figure 6 shows the evolution of the normalized L2 norm of the cost functional gradient during the iterative process. Figure 7 shows the evolution of the normalized L2 difference norm between the solution and the design profiles. Figure 8 shows the evolution of the global percentage of error during the minimization process.

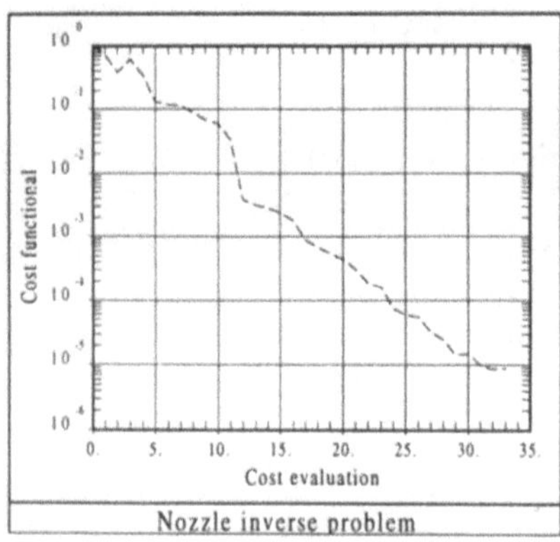

Figure 5. Evolution of the normalized cost functional.

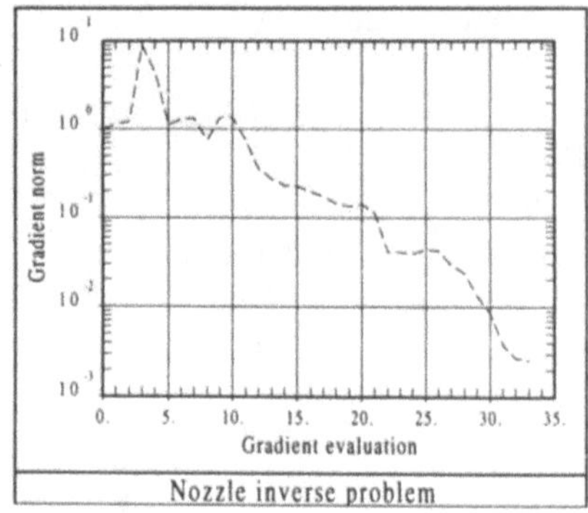

Figure 6. Evolution of the normalized L2 norm of the gradient of the cost functional.

Figures 5,6 and 7 show the fast convergence of the minimization process. The cost functional has been reduced 5 orders of magnitude in only 33 iterations. The norm of the objective function gradient has been reduced 4 orders of magnitude and the L2 difference norm has been reduced 3 orders of magnitude.

The global level of the error involved in the finite element computations has been fully controlled. After the seventh iteration it has been kept below the 0.1% limit imposed at the onset.

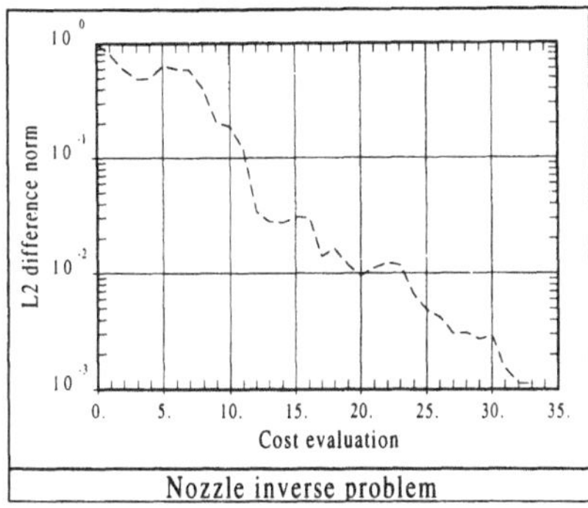

Figure 7. Evolution of the normalized L2 difference norm .

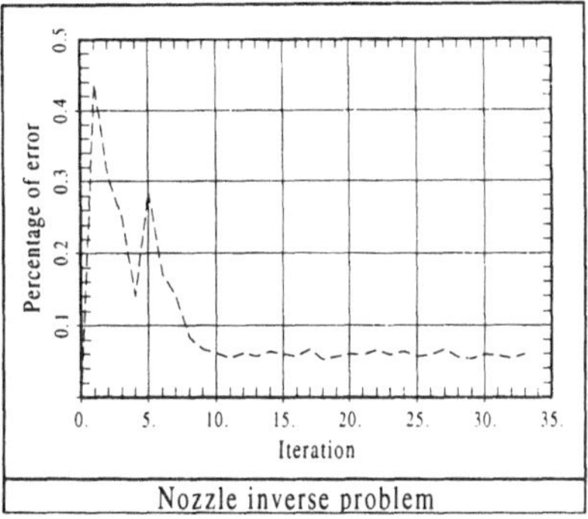

Figure 8. Evolution of the percentage of error.

9.2 T2 workshop test case

The T2 workshop test case consists in finding the shape which minimizes a cost function defined by the following integral expression over the bottom line of the nozzle:

$$f = \int_0^2 \left| \frac{\partial p}{\partial x} \right|_{y=0}^{\alpha} dx, \qquad \alpha = 2 \;\; or \;\; 4 \tag{27}$$

with the constraint $y(1) = 0.25$

The geometry of each design has been defined using 6 design variables. These variables are the y coordinates of 6 points which are used to interpolate a B-spline along the top line. Figure 9 shows the initial shape, the position of the 6 design variables and the finite element mesh used for the initial design. This initial design is the target design defined for the T1 test case. The constraint $y(1) = 0.25$ is automatically achieved because the coordinates of the corresponding point are fixed. The maximum global error has been limited to 0.1% of the total potential norm.

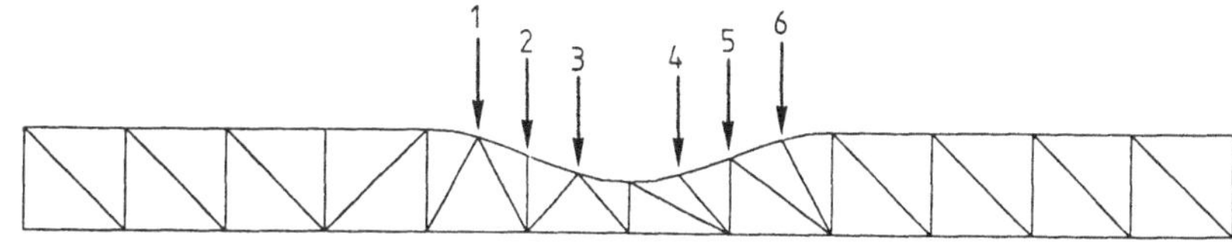

Figure 9. Initial shape, initial mesh and definition of the design variables.

This problem has been solved using $\alpha = 2$ (case T2.1) and $\alpha = 4$ (case T2.2) in both cases a strange solution has been obtained.

The final shape and the final mesh for case T2.1 can be observed in Figure 10. This shape has been obtained after 40 iterations. At this stage the process is not converged because the height of the waves seems to grow indefinitely. On the other hand, the value of the cost function remains almost constant during the last 20 iterations.

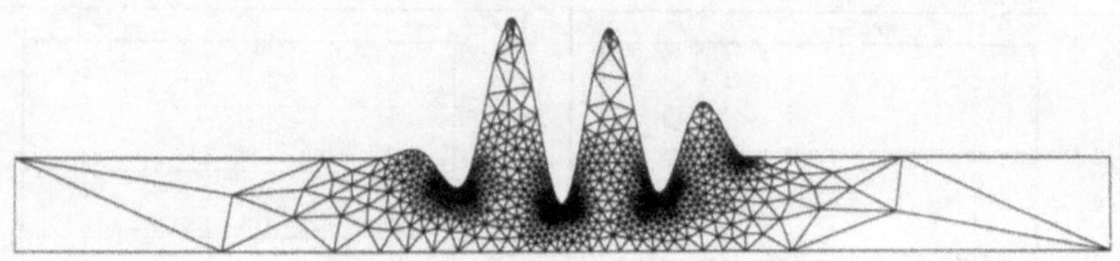

Figure 10. Final shape and final mesh for case T2.1 (f from (27) and $\alpha = 2$).

The final shape and the final mesh for case T2.2 ($\alpha = 4$) can be seen in Figure 11. In this case the situation is very similar to case 1 but the formed waves are smaller.

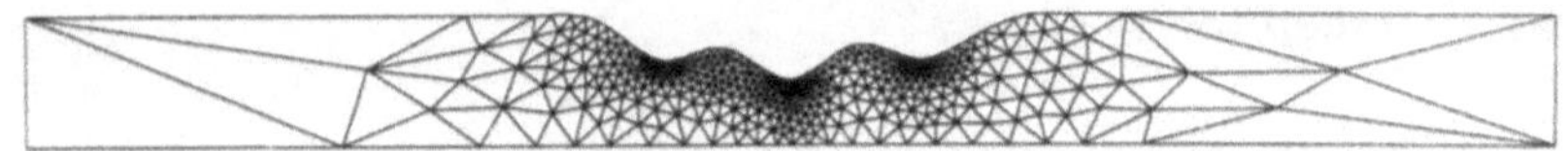

Figure 11. Final shape and final mesh for case T2.2 (f from (27) and $\alpha = 4$).

After some study of the results obtained for cases T2.1 and T2.2 a different cost function has been defined. This new cost function has the same expression than the initial one but the integral expression is extended over the top line of the nozzle, i.e.

$$f = \int_0^2 |\frac{\partial p}{\partial x}|^\alpha|_{y=y_{design}(x)} dx, \qquad \alpha = 2 \;\; or \;\; 4\,. \tag{28}$$

The final shape for this last case (case T2.3) can be observed in Figure 12.

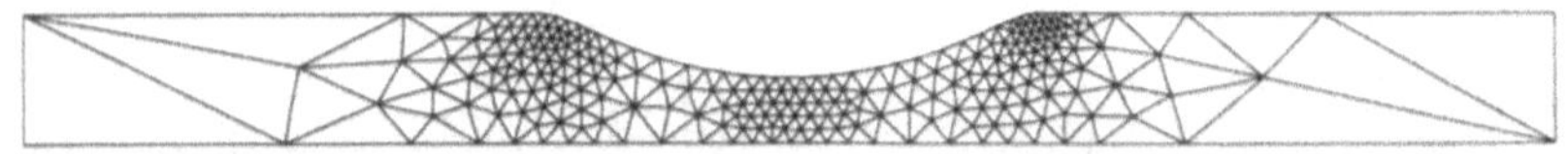

Figure 12. Final shape and final mesh for case T2.3 (f from (28) and $\alpha = 2$).

The iterative process has now converged after 24 iterations. The whole problem has taken around 10 minutes of CPU on a Personal Iris 35TG workstation.

Figure 13 shows the evolution of the normalized cost functional during the iterative process. Figure 14 shows the evolution of the normalized L2 norm of the cost functional gradient during the iterative process. Figure 15 shows the evolution of the global error during the minimization process.

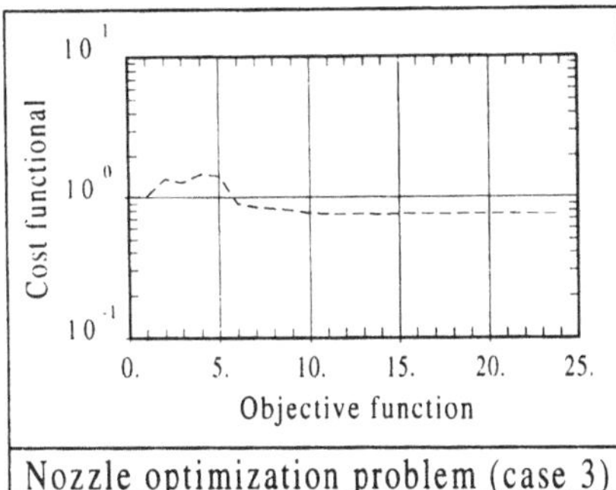

Figure 13. Evolution of the normalized cost functional.

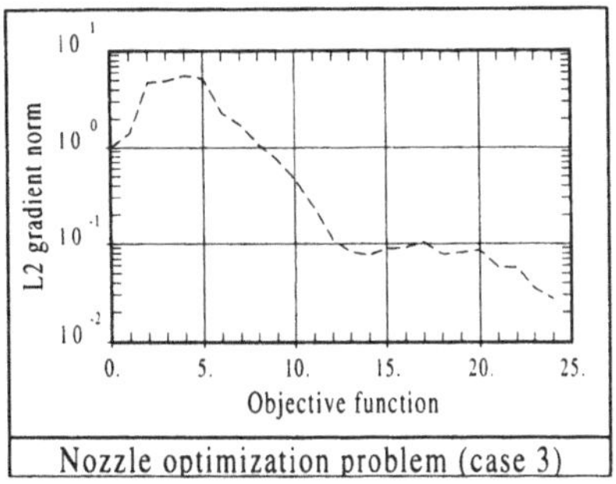

Figure 14. Evolution of the normalized L2 norm of the gradient of the cost functional.

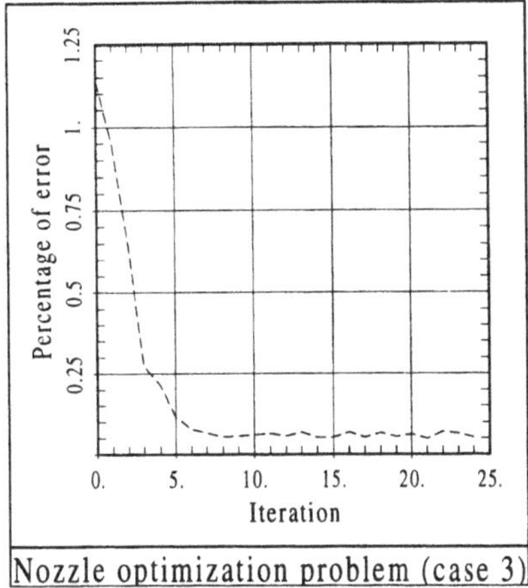

Figure 15. Evolution of the percentage of error.

The resolution of this test case has shown the importance of a good problem definition. A badly defined cost function can produce very strange final solutions, even if the process seems to converge very well.

Figures 13 and 14 show the convergence of the minimization process for case T2.3. The global error involved in the finite element computations has also been controlled and after the sixth iteration it has been kept below the 0.1% limit imposed at the beginning. Figure 12 shows a very satisfactory mesh for the final design.

9.3 T4 workshop test case

This test case consists in recovering the Korn airfoil at an angle of attack 0°.

The target pressure coefficient C_p^{target} has been obtained by a direct computation of the Korn airfoil with a finite element code using the same methodology as in test case T1 (section 9.1) and a maximum global error of 0.1%. The infinite boundary is placed a distance of 10 chords from the profile. The initial design corresponds to a NACA 64A410 profile.

The inverse problem has been solved using a minimization approach. The cost functional to be minimized has been defined as:

$$f = \int_0^2 (C_p(x) - C_p^{target})^2 ds \ . \tag{29}$$

This integral is extended around the profile and the integration variable is the arc s and not the x coordinate so that all the boundary is equally weighted. If the x variable is used the cost function tends to put more weight on the medium part of the profile and less on the edges.

The geometry of each design has been defined using 25 design variables. These variables are the y coordinates of 25 points distributed around the profile which are used to interpolate a B-spline. Figure 16 shows the initial shape and the finite element mesh used for the initial design. The 25 points used to define the shape of each design are all the nodes lying on the profile in Figure 16 with the exception of the trailing edge which is fixed. The maximum global error has been limited to 0.1% of the total potential norm.

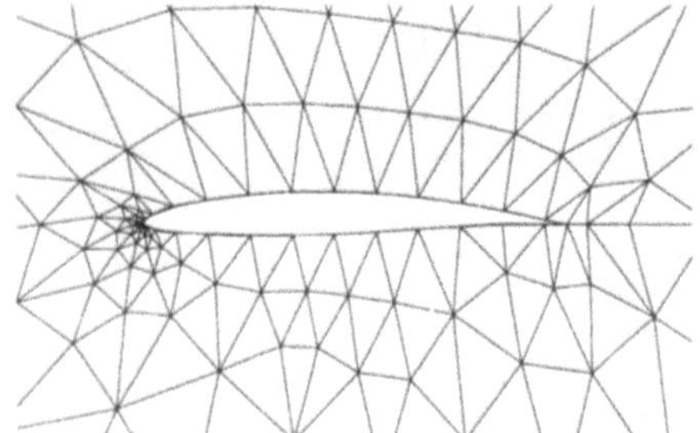

Figure 16. Initial shape and mesh.

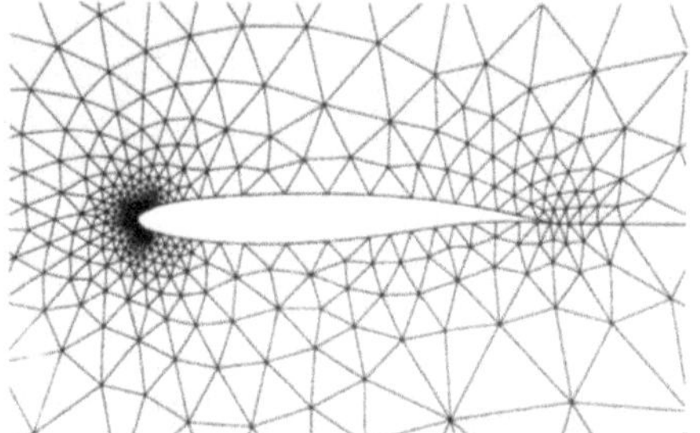

Figure 17. Finals shape and mesh.

The iterative process has converged after 50 iterations. The final shape and the final mesh can be observed in Figure 17. The whole problem has taken around 3.5 hours of CPU on a Silicon Graphics Indigo R4000 workstation.

The evolution of the normalized cost functional during the process can be seen in Figure 18. Figure 19 shows the evolution of the L2 difference norm between the solution profile and the design profile. This norm has been computed as the L2 difference norm between each design and the final one since no information on the exact definition of the Korn airfoil using the 25 control points was available. In fact it is not possible to define exactly the Korn airfoil using the 25 interpolating points. Figure 20 shows the evolution of the global error during the minimization process. The meshes used for the computations have around 2700 nodes and 1300 quadratic elements. Figure 21 shows the C_p distributions for the initial profile, the target profile and the last design obtained.

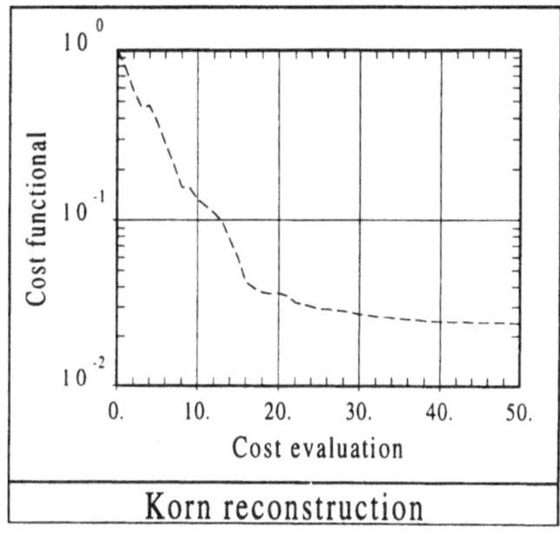

Figure 18. Evolution of the normalized cost functional during the iterative process.

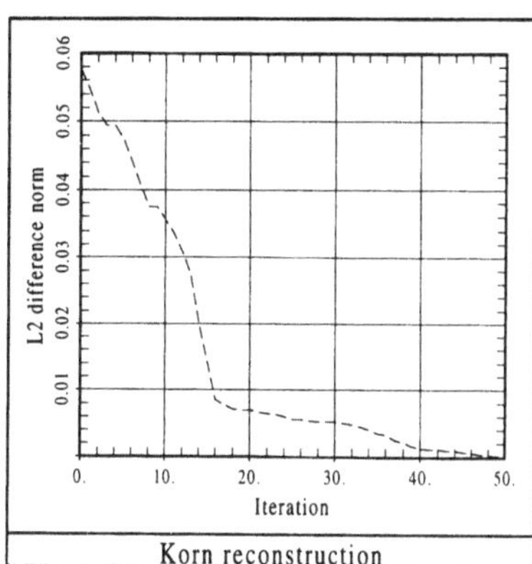

Figure 19. Evolution of the L2 difference norm between the solution profile and the design profile.

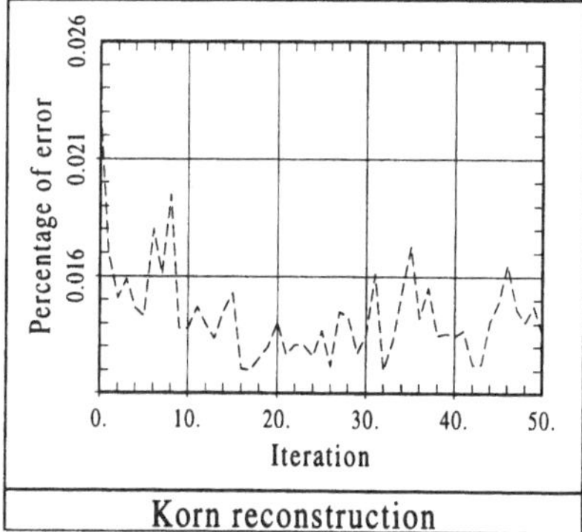

Figure 20. Evolution of the percentage of error.

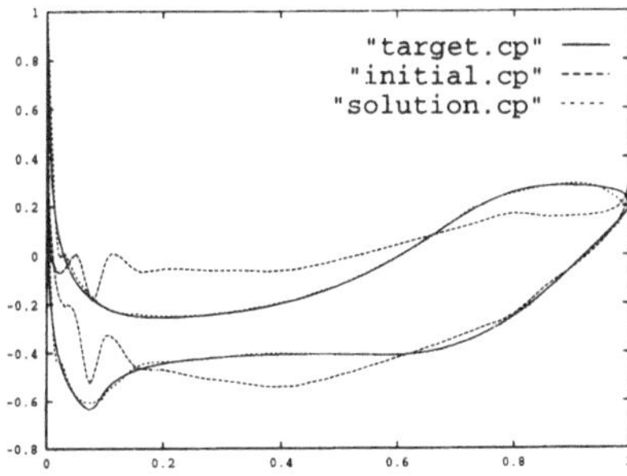

Figure 21. C_p distribution for the initial, target and computed designs.

Figures 18 and 19 show a good convergence of the minimization process. The cost functional has been reduced almost 2 orders of magnitude in 50 iterations.

The global error involved in the finite element computations has been controlled. In fact it is very low compared with the 0.1% limitation. This is because the error is concentrated around the profile, but a little bit far away the flow is almost uniform and the error is almost null. This explains why the global error is small. The important issue, in fact, is how the error is distributed around the profile. Figure 17 shows how the mesh concentrates many more elements around the leading edge where the gradients of the potential are higher.

The comparison between the target and the computed C_p values shown in Figure 21 is quite good although some differences are still noticeable. On the other hand Figure 18 shows that the process seems to be converged, and it is not possible to get a solution closer to the target one. The reasons are that probably it is not possible to get a better definition of the Korn airfoil using a B-splines interpolation with 25 points. In fact the computations over the Korn airfoil have shown that it is extremely sensible to little changes in its shape. In order to get a better final solution it would be necessary to use more design variables to enhance the B-spline interpolation, but this would considerable increase the total cost.

The conclusions from this test case are:

- The methodology used works in a very satisfactory way producing a good final solution with an adequate final mesh.
- The convergence of the process can also be considered good.
- A study of the parametrization required to obtain a solution closer to the target one appears as very necessary.

9.4 T10 workshop test case (preliminary work)

As a preliminary work to solve the T10 test case a problem with two airfoils have been studied. This case consists in recovering two NACA 0012 profiles at an angle of attack 0° positioned as shown in Figure 23. The starting design is formed by the two profiles as shown in Figure 22. Each profile has been obtained from a NACA 0012 reducing its thickness to one half and rotating each one 5° around their leading edges.

The target pressure coefficient C_p^{target} has been obtained by a direct computation of the target design as in the previous T4 test case.

The inverse problem has been solved using a minimization approach. The cost functional to be minimized has been defined as in eq. (29) extending the integral to the boundarys of both airfoils.

The geometry of each airfoil has been defined using 18 design variables. This variables are the y coordinates of 18 points distributed around each profile which are used to interpolate a B-spline. Figure 22 shows the initial shape and the finite element mesh used for the initial design. The 18 points used to define the shape of each profile are all the nodes lying on them in Figure 22 with the exception of the trailing and leading edges. For each profile, its rotation around its leading edge has been defined as an additional design variable. This means that two angles have been defined as design variables. The total number of design variables is thus 38. The maximum global error during the minimization process has been limited to a 0.2% of the total potential norm.

The iterative process has been considered as converged after 100 iterations. The final shape and the final mesh can be observed in Figure 23. The whole problem has taken around 100 hours of CPU on a CONVEX C-3480 computer using 1 single processor. It is important to notice that the code has not been adapted to take advantage of the vectorial capabilities of the computer which could substantially reduce the computational cost.

The evolution of the normalized cost functional can be seen in Figure 24. Figure 25 shows the evolution of the normalized L2 difference norm between the solution profile and the design profile. Figure 26 shows the evolution of the global error during the minimization process. The meshes used for the computations have around 2700 nodes and 1300 quadratic elements. Figures 27 and 28 show the superposition of the C_p distribution for the target profile and the last design for each airfoil.

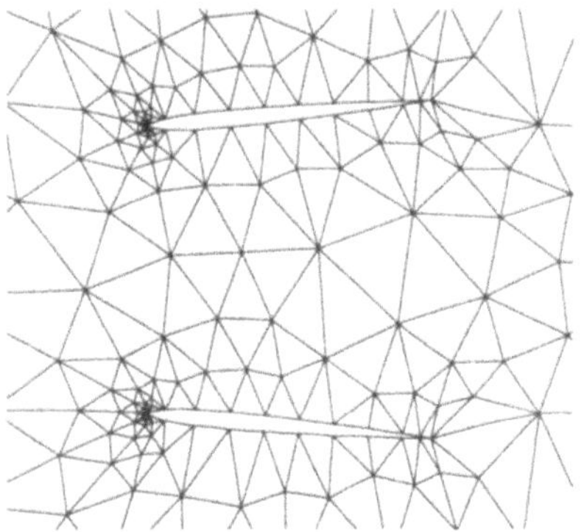

Figure 22. Initial shape and mesh.

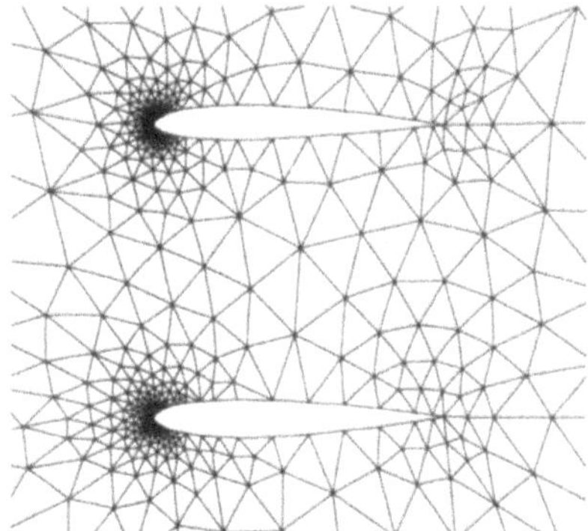

Figure 23. Finals shape and mesh.

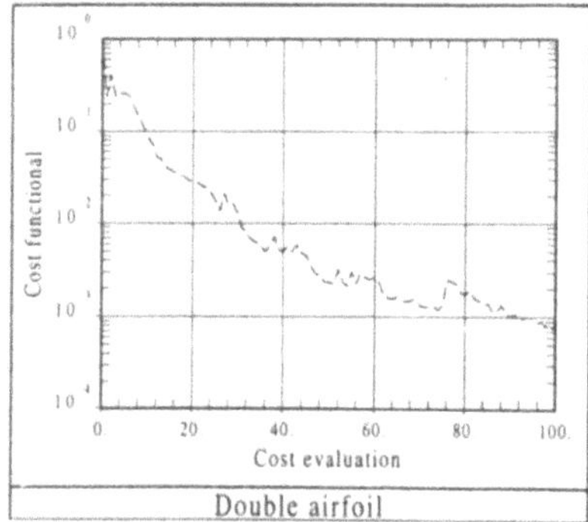

Figure 24. Evolution of the normalized cost functional.

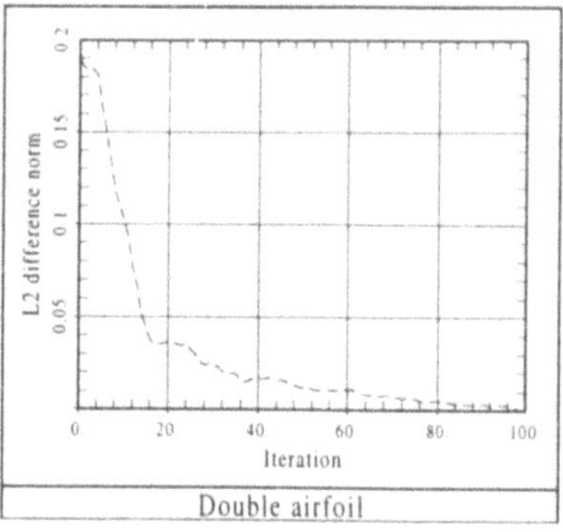

Figure 25. Evolution of the normalized L2 difference norm.

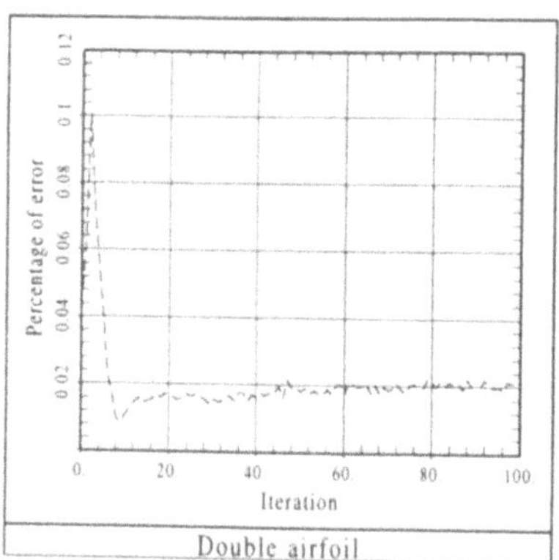

Figure 26. Evolution of the percentage of error.

Figures 24 and 25 show a good convergence of the minimization process. The cost functional has been diminished more than 3 orders of magnitude in 100 iterations.

The global error involved in the finite element computations has been completely

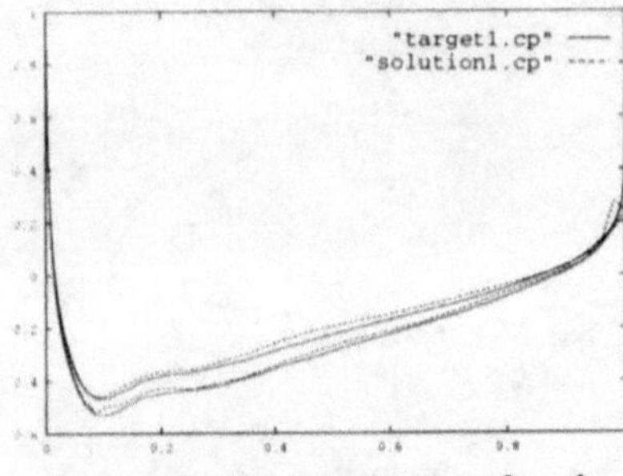

Figure 27. C_p distribution for the target and computed upper airfoils.

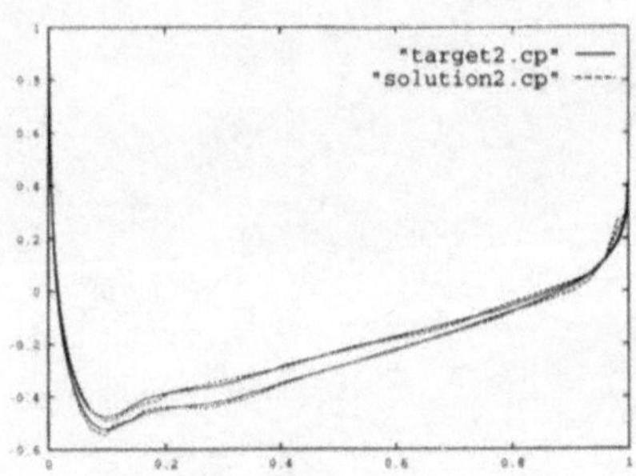

Figure 28. C_p distribution for the target and computed lower airfoils.

controlled. In fact, as in the previous T4 test case, it is very low compared with the imposed limitation 0.2%.

The comparison between the target and the computed C_p shown in Figures 27 and 28 is quite good.

This example shows that the present methodology is able to solve a problem similar to the T10 workshop test case with more than one airfoil. On the other side a direct computation of the flow around the target profile of the T10 test case produces a solution with a pressure distribution very close to the target one proposed for the workshop. Nevertheless the T10 problem has not been solved due to convergence problems of the process. Further research is being developed in order to solve this problem and to present the corresponding results in the second edition of the workshop about Optimum Design in Aerodynamic problems.

10. CONCLUSIONS

A new methodology for the resolution of optimization and inverse problems has been developed and assessed. This methodology is able to optimize the design and the analysis mesh together in order to produce a final design computed with a proper mesh.

Good quality results are obtained using a single mesh for each design without any remeshing. This considerably reduces the additional cost of the mesh control.

The presented methodology has provided excellent results for all the test cases analyzed leading to an accurate final solution with a good final mesh.

The use of a different "optimal" mesh for each design seems to be specially interesting to apply this methodology to more realistic flow models where the control of the mesh is crucial. This could be particularly attractive in presence of shocks.

11. ACKNOWLEDGMENTS

The authors acknowledge the support of the Commission of the European Communities DG-XII through the BRITE/EURAM project "Optimum Design in Aerodynamics".

12. REFERENCES

[1] Bugeda, G. "Utilización de técnicas de estimación de error y generación automática de mallas en procesos de optimización estructural." Ph. D. Thesis - Universitat Politècnica de Catalunya (1990) (In Spanish).

[2] Bugeda, G. and Oñate, E., "New adaptive techniques for structural problems", First European Conference on *Numerical Methods in Engineering*, Brussels, Belgium, September, 1992.

[3] Bugeda, G. and Oñate, E., "Adaptive mesh refinement techniques for aerodynamic problems", in *Numerical Methods in Engineering and Applied Sciences.* H. Alder, J. C. Heinrich, S. Lavanchy, E Oñate & B. Suárez (Eds.), CIMNE-Barcelona 1992.

[4] Bugeda, G., Oñate, E., and Miquel, J. "Optimum Design in Aerodynamics" BRITE/EURAM AREA 5, Contract N° AERO-0026C, Proposal N° 1082. Mid-Term Report (1991).

[5] Bugeda, G., Oñate, E., and Miquel, J. "Optimum Design in Aerodynamics" BRITE/EURAM AREA 5, Contract N° AERO-0026C, Proposal N° 1082. Final Report (1992).

[6] Faux, I. D. and Pratt, M. J. "Computational Geometry for Design and Manufacture.", Edited by Ellis Horwood Limited, 1987.

[7] Navarrina, F. "Una metodología general para optimización estructural en diseño asistido por ordenador." Ph. D. Thesis - Universitat Politècnica de Catalunya (1987) (In Spanish).

[8] Navarrina, F., Bendito, E. and Casteleiro M. "High order sensitivity analysis in shape optimization problems." - *Computer Methods in Applied Mechanics and Engineering, vol. 75, pp. 267-281 (1989)*

[9] Oñate, E., Castro, J. and Kreiner, R., "Error estimations and mesh adaptivity techniques for plate and shell problems", presented at the *3rd. International Conference on Quality Assurance and Standards in Finite Element Methods*, Stratford-upon-Avon, England, 10–12 September, 1991.

[10] Oñate, E. and Castro, J., "Adaptive mesh refinement techniques for structural problems", published in *"The Finite Element Method in the 90's. A book dedicated to O. C. Zienkiewicz"*, E. Oñate, J. Periaux and J. Samuelsson (Eds.), Springer-Verlag&CIMNE, Barcelona 1991.

[11] Oñate, E. and Bugeda, G., "A Study of Mesh Optimality Criteria in Adaptive Finite Element Analysis" *"Engineering Computations" Vol. 10, Num. 4, (1993)*

[12] Peraire, J. "A Finite Element Method for Convection Dominated Flows." - Ph. D. Thesis - University College of Swansea (1986)

[13] Peraire, J., Morgan, K. and Peiró, J. "Unstructured finite element mesh generation and adaptive procedures for CFD." - *AGARD FDP: Specialist's Meeting, Loen, Norway (1989)*

[14] Pironneau, O. "Méthodes des éléments finis pour les fluides." Masson 1988.

[15] Zienkiewicz, O.C. and Zhu, J.Z., "A simple error estimator and adaptive procedure for practical engineering analysis", *Int. Num. Meth. Engrg.*, **24**, 337–357, 1987.

[16] Zienkiewicz, O.C., Zhu, J.Z., Liu, I.C., Morgan, K. and Peraire, J., "Error estimates and adaptivity from elasticity to high speed compressible flow", in J.R. Whiteman, ed., MAFELAP 87, 483–512, Academic Press, New York, 1988.

[17] Zhu, J.Z. and Zienkiewicz, O.C., "Adaptive techniques in the finite element method", *Comm. Appl. Numer. Methods*, **4**, 197–204, 1988.

[18] Zienkiewicz, O.C., Zhu, J.Z. and Gong, N.G., "Effective and practical $h-p$ version adaptive analysis procedures for the finite element method", *Internat. J. Numer. Methods Engrg.*, **28**, 879–891, 1989.

[19] Brite Euram 1082 partners, "Workshop on Selected Inverse and Optimum Design Problems. Definition of Test Cases" (1992).

SINGLE-PASS METHOD FOR THE SOLUTION OF THE 2-D & 3-D INVERSE POTENTIAL PROBLEM

V. Dedoussis, P. Chaviaropoulos, K.D. Papailiou

National Technical University of Athens
Lab. of Thermal Turbomachines
P.O. Box 64069, 157 10 Athens
Greece.

SUMMARY

A method for the solution of the 2-D and 3-D inverse potential problem is presented. Potential and streamfunction are introduced in order to map the physical space onto a computational one via a body-fitted coordinate transformation. The velocity magnitude, the aspect ratio and the cross-section angle of the elementary streamtubes are the dependent variables. A novel procedure based on differential geometry and generalized tensor analysis arguments is employed to formulate the method and derive the governing differential equations. The assumption of orthogonal streamsurfaces reduces the number of dependent variables by one, simplifying the governing equations to an elliptic p.d.e. for the velocity magnitude and a second order o.d.e. for the streamtube aspect ratio. The solution of these two equations provides the flow field. Geometry is determined independently by integrating Frenet equations of the grid lines, after the flow field has been determined. The 2-D case is treated as a particular case of the general 3-D one. The method has been applied to both 2-D and 3-D reproduction cases of the present workshop with very satisfactory results.

INTRODUCTION

The development of reliable methods for designing efficient aerodynamic components is a subject of great importance for the aeronautics community. In most of the cases the performance of these components depends on the characteristics of the boundary layers developing along their walls, which in turn depend heavily on the wall pressure distribution. The designer therefore has to tackle the problem of determining the shape of the walls on which the pressure (or velocity in inviscid flows) distribution is prescribed.

Various methods have been developed which solve the above "target pressure" problem. First attempts in the field are traced back in the forties, when conformal mapping methods [1] were used for the solution of the inverse problem of two-dimensional (2-D) incompressible potential flows. To the credit of these methods was the design of several airfoil series with relatively low drag compared to the older, empirically designed NACA series. In the early fifties, Stanitz [2] developed an inverse method, initially for incompressible and later for compressible 2-D flows. His idea was to solve the governing equations on a transformed domain employing the potential function (ϕ) and the streamfunction (ψ) as "natural coordinates".

In recent years, optimization methods have been developed in an attempt to circumvent the inherent limitations of the above methods which were basically 2-D and irrotational. In these methods, the design geometry is determined via a series of analysis problems. Their major advantage is the flexibility in the definition of the cost-function to be optimized (the target pressure problem being one of the variants) as well as the possibility of incorporating geometrical or "flow" constraints. The disadvantage of the optimization methods is that, for the time being, they are time consuming (a few hundreds of analysis iterations may be required plus the regriding overhead). Reviews on aerodynamic shape design methods have been published by Dulikravich [3] and recently by Labrujere and Slooff [4].

The purpose of this paper is to present a fast, single-pass, potential compressible inverse method for the design of 2-D and 3-D configurations. This method has been developed at the Lab. of Thermal Turbomachines of the National Technical University of Athens in the context of the EEC BRITE EURAM AERO-0026-C(TT) "Optimum Design in Aerodynamics" Project.

Similar to the method proposed by Stanitz [2] [5] [6], the present method introduces a potential function (ϕ) and two streamfunctions (ψ,η) as the "natural coordinates". The physical (x,y,z) space, on which the boundaries of the flow field sought are unknown, is mapped onto the potential-streamfunctions (ϕ,ψ,η) space via a body-fitted coordinate transformation. Computational boundaries on the latter space are fixed simply because, in inviscid flows, solid walls are stream surfaces, i.e. ψ=const. or η=const. surfaces, and inlet and outlet sections can be considered to be potential ones. Thus assuming that the velocity distribution -prescribed pressure- is given on the walls, as well as on the inlet and outlet sections of the channel, one is faced, with solving a boundary value problem on the (ϕ,ψ,η) space.

The present method has been developed in a mathematically formal way, employing differential geometry and generalized tensor analysis arguments. Both the formulation and the governing equations are quite different from those proposed by Stanitz. The equation for the velocity magnitude is supplemented by another equation for the aspect ratio of the cross-section of the elementary streamtube. These two equations provide the flow solution. The present method is a fast one, since the flow solution does not depend on the geometry. The geometry is determined (by integrating Frenet equations on the grid lines) after the flow solution has been found. Another advantage is that the present approach enables the proper investigation of the existence and uniqueness of solution, of the general 3-D inverse problem [7], [8].

The 2-D case is treated as a particular case of the general 3-D one. In the 2-D case, however, the aspect ratio of the streamtubes is directly related to the velocity logarithm and therefore the corresponding equation becomes redundant. In external aerodynamic cases, special care has been taken in order to handle the so called "leading edge" singularity problem [9].

The method has been applied to two reproduction test cases for 2-D internal and external flows and a 3-D internal one, which have been defined in the present workshop. The satisfactory results included in this paper indicate the accuracy of the proposed method.

BASIC ASSUMPTIONS AND EQUATIONS

In the development of the present inverse method it is assumed that the flow is 3-D, steady, subsonic and irrotational. It is also assumed that the fluid is a perfect gas. In accordance with the above mentioned assumptions the basic flow equations are:

Continuity equation

$$\nabla\cdot(\rho\vec{V})=0\,. \tag{1}$$

Irrotationality condition

$$\nabla\times\vec{V}=0\,. \tag{2}$$

Density equation (energy conservation for isentropic changes)

$$\rho=\left[1+\frac{\gamma-1}{2}M_\infty^2\,(1-V^2)\right]^{1/(\gamma-1)}. \tag{3}$$

In the above equations the velocity (V) is normalized with a reference value (V_∞) and the density (ϱ) with the corresponding (ϱ_∞) value. M_∞ is the Mach number at the reference point. γ is the ratio of specific heats c_p/c_v.

THREE DIMENSIONAL METHOD FORMULATION

Potential function

The irrotationality condition of the velocity field expressed by equation (2) is satisfied identically, by requiring the velocity vector to be the gradient of a scalar function, i.e. potential function. The potential function ϕ is defined by the relation:

$$\vec{V}=\nabla\phi\,. \tag{4}$$

Streamfunctions

Two streamfunctions (streamvectors) ψ,η are introduced. They are defined by the relation:

$$\rho\vec{V}=\nabla\psi\times\nabla\eta\,. \tag{5}$$

It is evident that the two streamfunctions are defined in such a way so to satisfy the continuity equation (1) identically. Equation (5) indicates that the velocity vector is tangent to both ψ=const. and η=const. surfaces, which are appropriately termed as stream surfaces. Obviously intersections of stream surfaces, which belong to a different family, are streamlines. Schematically potential and stream surfaces are shown in Fig.1.

Natural curvilinear (ϕ,ψ,η) coordinate system

The potential function ϕ and the two streamfunctions ψ,η are considered to be the independent variables. The contravariant base of the curvilinear (ϕ,ψ,η) coordinate system is:

$$\vec{g}^1=\nabla\phi \quad , \quad \vec{g}^2=\nabla\psi \quad , \quad \vec{g}^3=\nabla\eta \tag{6}$$

where coordinate indices 1,2,3, i.e. coordinates x^1,x^2,x^3, are associated with the ϕ,ψ,η coordinates respectively.

The elementary distance ds expressed in terms of the natural coordinates (ϕ,ψ,η) is:

$$ds^2=g_{11}d\phi^2+g_{22}d\psi^2+g_{33}d\eta^2+2g_{23}d\psi\, d\eta \tag{7}$$

where the metrics g_{ij} are expressed in terms of the flow quantities V and ϱ, and two geometrical dependent variables θ and t. Variable θ is the angle at which a ψ=const. surface intersects a η=const. surface on a potential, ϕ=const. surface (i.e. the angle between $\vec{g}_2$ and $\vec{g}_3$ in Fig.1), whilst variable t is defined as:

$$t^2=\frac{g_{33}}{g_{22}} \tag{8}$$

representing the aspect ratio of the elementary streamtube. The diagonal elements of the metric tensor for instance are:

$$g_{11}=\frac{1}{V^2} \quad , \quad g_{22}=\frac{1}{\rho V t \sin\theta} \quad , \quad g_{33}=\frac{t}{\rho V \sin\theta} \,. \tag{9}$$

Complete expressions of the metrics and conjugate metrics tensors, g_{ij} and g^{ij} respectively, are given in [7] and [8].

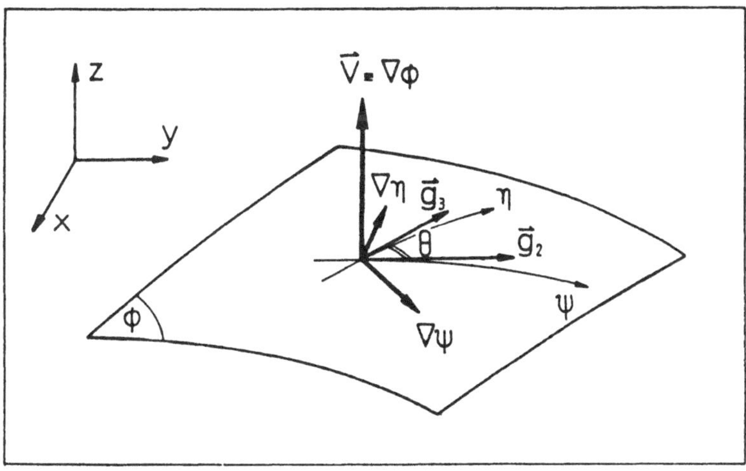

Figure 1. Natural (ϕ,ψ,η) and physical (x,y,z) coordinate systems.

GOVERNING EQUATIONS

Differential geometry and generalized tensor analysis arguments are employed in order to derive the governing equations. It is assumed that the 3-D (x,y,z) Euclidean space is described by the natural curvilinear (ϕ,ψ,η) coordinate system. Euclidean space being a flat space has zero curvature. Referring to the Ricci curvature tensor [10] the zero curvature condition reads

$$R_{rm}=0 \qquad with\ r,m=1,2,3. \tag{10}$$

Using generalized tensor notation the Ricci tensor is defined as:

$$R_{rm}=\frac{\partial\Gamma^{n}_{rn}}{\partial x^{m}}-\frac{\partial\Gamma^{n}_{rm}}{\partial x^{n}}+\Gamma^{p}_{rn}\,\Gamma^{n}_{pm}-\Gamma^{p}_{rm}\,\Gamma^{n}_{pn} \tag{11}$$

where

$$\Gamma^{p}_{rn}=g^{pl}\,[rn,l] \tag{12a}$$

$$[rn,l]=\frac{1}{2}\left[\frac{\partial g_{rl}}{\partial x^{n}}+\frac{\partial g_{nl}}{\partial x^{r}}-\frac{\partial g_{rn}}{\partial x^{l}}\right] \qquad with\ l,m,n,p,r=1,2,3. \tag{12b}$$

Relations (12a) and (12b) define the Christoffel symbols of the second and first kind respectively.

Since the metrics of the (ϕ,ψ,η) coordinate system are expressed in terms of the flow quantities V,ϱ and the characteristics of the elementary streamtubes t, θ (refer to relations (9)), it is evident, that the zero curvature conditions (10), can provide p.d.es in terms of the (dependent) variables V,ϱ,t,θ.

The governing p.d.es for V, t and θ are derived by the following combinations of the zero conditions for the elements of the Ricci tensor:

Velocity (V) equation

$$-R_{11}/g_{11}=0. \tag{13}$$

Aspect Ratio (t) equation

$$-R_{22}/g_{22}+R_{33}/g_{33}=0. \tag{14}$$

Coordinate Angle (θ) equation

$$R_{22}/g_{22}+R_{33}/g_{33}-2R_{23}/g_{23}-R_{11}/g_{11}=0. \tag{15}$$

Equations (13)-(15) supplemented by equation (3), constitute a closed set of p.d.es for the quantities V,ϱ,t,θ, in terms of the ϕ,ψ,η independent variables.

In the general case where θ is a genuine variable it has been shown [8] that the set of the governing equations (13)-(15) accepts multiple solutions when boundary conditions are prescribed for the velocity field only. However, for $\theta = 90^o$ equation (15) is satisfied identically, indicating implicitly, that the flow with rectangular sectioned elementary streamtubes constitutes a partial solution of the 3-D inverse problem. Taking advantage of this property the relevant V and t equations, (13) and (14) respectively, read:

V-equation

$$
\begin{aligned}
&V[(lnV)_{\phi\phi} + (ln\rho)_{\phi\phi}] + \rho t (lnV)_{\psi\psi} + \frac{\rho}{t}(lnV)_{\eta\eta} + \\
&+ \frac{1}{2} V[(lnV)^2_{\phi} - (lnt)^2_{\phi} - (ln\rho)^2_{\phi}] - \rho t (lnV)_{\psi}[(lnV)_{\psi} - (lnt)_{\psi}] - \\
&- \frac{\rho}{t}(lnV)_{\eta}[(lnV)_{\eta} + (lnt)_{\eta}] = 0
\end{aligned}
\tag{16}
$$

t-equation

$$
\begin{aligned}
&V[(lnt)_{\phi\phi} - (ln\rho)_{\phi}(lnt)_{\phi}] + \rho t[(lnV)_{\psi\psi} + (lnV)_{\psi}(ln\rho)_{\psi}] - \\
&- \frac{\rho}{t}[(lnV)_{\eta\eta} + (lnV)_{\eta}(ln\rho)_{\eta}] = 0
\end{aligned}
\tag{17}
$$

where subscripts ϕ,ψ,η indicate corresponding partial derivatives. Equation (3) is used to evaluate the density from the velocity field.

The governing equations (16) and (17) are integrated on a parallepiped domain in the (ϕ,ψ,η) space. ϕ=const. surfaces correspond to the inlet and outlet sections, whilst ψ=const. and η=const. surfaces correspond to the lateral solid boundaries.

V-equation (16) being a second order elliptic p.d.e requires boundary conditions all round the integration domain. Velocity is specified (Dirichlet type condition) on the solid walls -target velocity distribution- and on the inlet and outlet sections. The latter two distributions are assumed to be uniform.

The potential ϕ is related to the arc length s of the limiting streamlines on the solid walls via the relation $d\phi = Vds$. It is obvious therefore, that $V=V(\phi)$ could be considered to represent a $V=V(s)$ distribution. The designer usually specifies the distribution $V=V(s)$ rather than $V=V(\phi)$.

t-equation (17) is a second order o.d.e for t, in the ϕ-wise sense. Uniform Dirichlet type boundary condition is used for the inlet section, while zero Neumann-type boundary condition is imposed at the exit section (i.e. fully developed flow conditions).

FLOW FIELD CALCULATION

The governing V- and t-equations are highly nonlinear and have different mathematical character. It has been decided therefore to solve them in a coupled iterative mode. The equations are first linearized using a Newton procedure. Second order accurate centered differencing on a staggered grid is employed for the discretization of the linearized governing equations. The resulting linearized algebraic system of equations is solved with the restarting linear GMRES (m) [11] procedure. Details of the numerical procedure can be found in [7] [12]. It is worthnoting that the flow field is provided in a self contained "single-pass" manner without requiring any feedback from the geometry. This characteristic renders the present method a fast one.

GEOMETRY CALCULATION

Having calculated the flow field on the (ϕ,ψ,η) space, the geometry of the channel is determined by integrating the Frenet equations on the grid lines of the natural (ϕ,ψ,η) space, on the physical (x,y,z) space. The 3-D form of the Frenet equations of the natural space grid lines are:

$$\frac{\partial \vec{g}_i}{\partial x^j}=\Gamma^k_{ij}\vec{g}_k \qquad \text{with } i,j,k=1,2,3 \tag{18}$$

where

$$\vec{g}_i=\frac{\partial \vec{r}}{\partial x^i} . \tag{19}$$

$\vec{r}$ is the position vector in the physical (x,y,z) Cartesian space. It is noted that the Christoffel symbols Γ_{ij}^k are expressed in terms of the partial derivatives of the calculated flow field quantities V,ρ,t (see equations (9),(12)).

Equations (18) represent a 9x9 system of o.d.es along each grid direction for the Cartesian components of the covariant base $(\vec{g}_1,\vec{g}_2,\vec{g}_3)$. Once the covariant base components become available at the grid nodes, integration of equation (19) provides the Cartesian coordinates of the grid nodes.

TWO-DIMENSIONAL CASE

The 2-D case is treated as a particular case of the general 3-D one. Considering that the x^3-wise direction is the normal to the plane of the flow direction, then g_{33} is identically equal to one, implicitly defining t, via eq.(9), as:

$$t=\rho V. \tag{20}$$

It can be shown, that for t given by equation (20), the two equations governing the V- and t-fields, (16) and (17) respectively, become identical with one another. This implies that, in the 2-D case, t-equation is redundant. The resulting single equation is given in [9], which is the same with the one used by Stanitz [2] in designing 2-D channels.

External aerodynamic case calls for special attention because of the "leading edge" singularity. In all previous efforts the H-type grid, which has been used for the discretization of the (ϕ,ψ) plane (in the sense that grid lines are aligned with the ϕ=const. and ψ=const. lines) performed poorly in the neighbourhood of the leading edge stagnation point. In the present work, an additional numerical transformation is carried out which maps a rectangular computational domain on a C-type grid on the (ϕ,ψ) plane. This transformation, which employs elliptic coordinates is a simplified version of the Rizzi [13] grid generator (see also [9]).

The relevant simpler 2-D version of the Frenet equations (18) and (19) of the potential and streamlines are used for determining the geometry of the channel/body in the 2-D case.

COMPUTATIONAL RESULTS AND DISCUSSION

The proposed inverse design method has been applied to the subsonic test cases T1, T4 and T14 of the present workshop.

T1 TEST CASE 2-D Subsonic Nozzle

Test case T1 concerns the reproduction of a 2-D subsonic symmetric convergent-divergent nozzle whose upper wall shape is prescribed by a sinusoidal curve. In order the bottom wall to be a straight line, a mirror image nozzle has been assumed. The computational domain has been extended upstream and downstream the "interesting" region. M_∞ has been set to 0.2.

A full-potential solver is used in order to obtain the wall velocity -target pressure-distribution. A simple 61×31 grid, incorporating straight lines in the y-sense, has been used. Wall velocity and the corresponding pressure coefficient (C_p) distributions calculated by the direct solver are shown in Fig.2.

A 61x31 square grid was generated on the (ϕ,ψ) plane. The computational cost associated with the inverse solution, i.e. geometry reproduction, is of the order of 8 CPU secs on an Alliant FX 80 computer. It is noted that the present method provides not only the geometry of the nozzle, but also a complete grid system with desirable properties (orthogonal and stretched in "interesting" areas). This grid is presented in Fig.3. Inverse and direct solvers Mach contours are compared in Fig.4. The agreement is very good. The difference between the calculated y-coordinate of the top wall of the nozzle and the corresponding target coordinate is presented in Fig.5. Differences are of the order of 10^{-4} with the value of the l^2-norm being $2.1\text{x}10^{-4}$. One may observe the kinks in the error distribution which are due to the discontinuity of the second order derivative of the nozzle boundary (junctions of the sinusoidal curve with the straight extension lines).

T4 TEST CASE Single Subsonic Airfoil

Test case T4 concerns the reproduction of the Korn airfoil under subsonic flow conditions. M_∞ is .75 and the angle of attack is 0 degrees.

A full-potential solver is used in order to obtain the suction and pressure side wall velocity distributions. A C-type grid with 165x21 nodes has been used for the flow field computation. This grid, which is generated using elliptic coordinates, redistributes the profile coordinates in order to obtain smooth derivatives representation near the sensitive leading and trailing edge regions. It is pointed out, however, that the Korn profile is very sensitive to its coordinates description and the redistribution process, therefore, may cause discrepancies to the flow field computation. The Mach number and isobar contours calculated by the direct solver are shown in Figs 6 and 7 respectively.

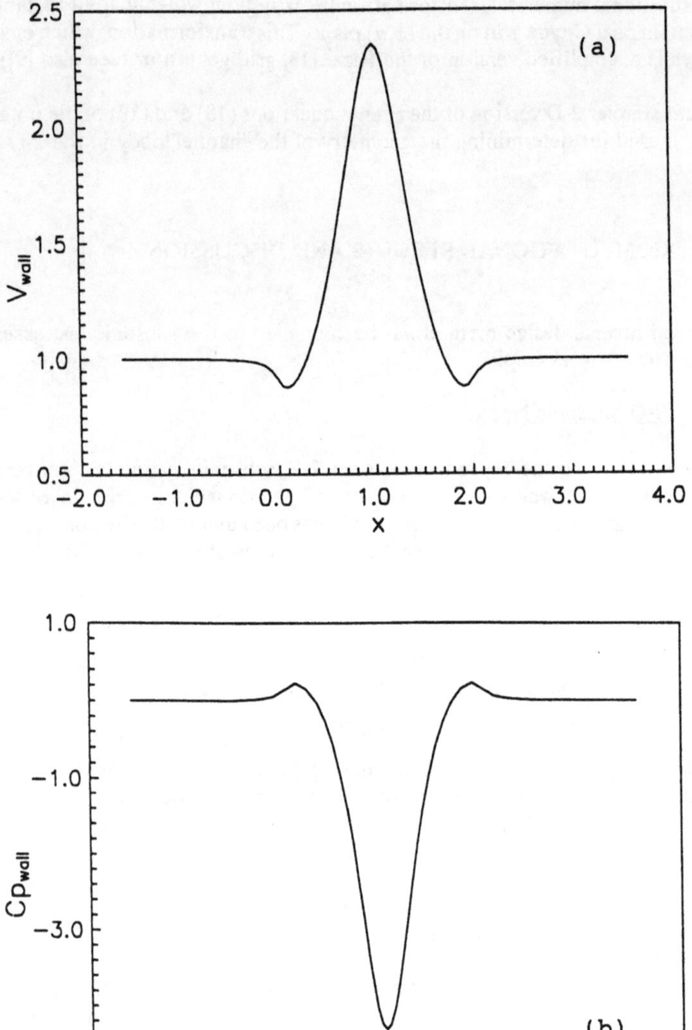

Figure 2. Upper wall distributions for (a) the velocity and (b) the pressure coefficient of the direct solver for the 2-D nozzle case.

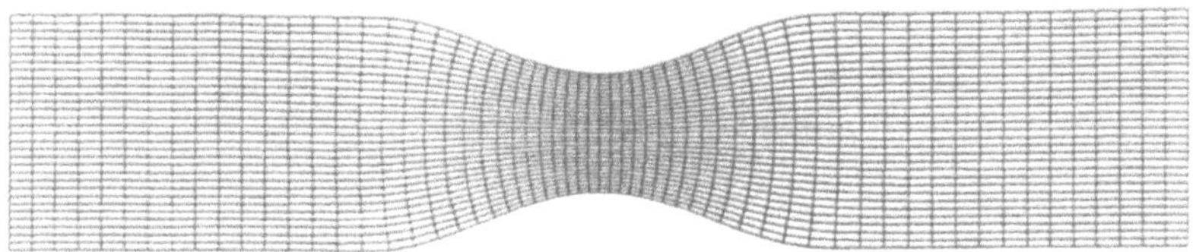

Figure 3. Geometry and grid calculated by the inverse method.

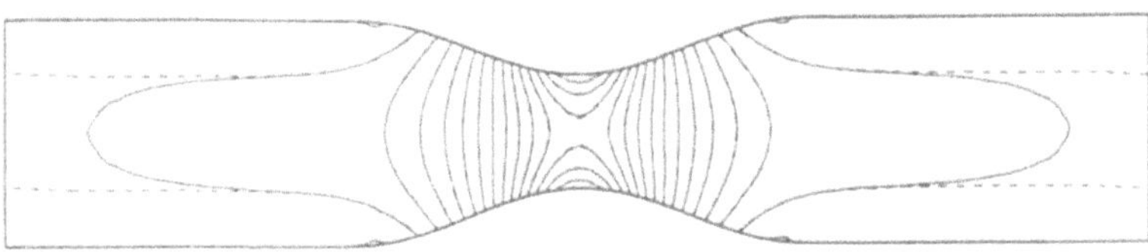

Figure 4. Mach contours of inverse (——) and direct (---) calculations (M_{min}=.18, ΔM=.02).

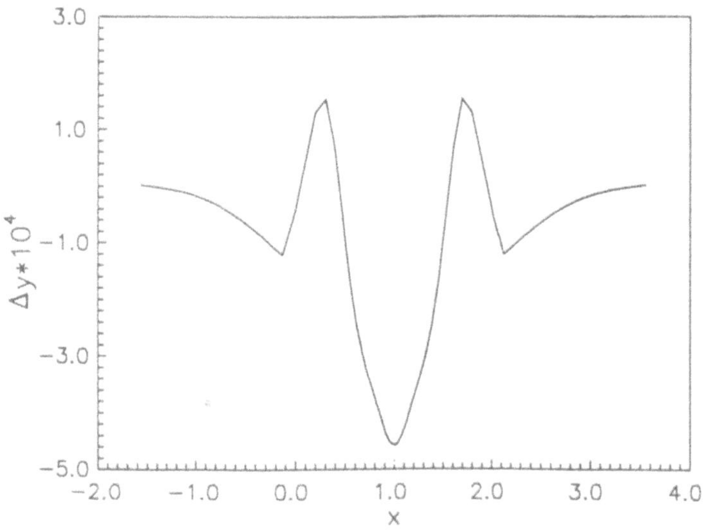

Figure 5. Distribution of calculated geometry error of the upper wall of the 2-D nozzle.

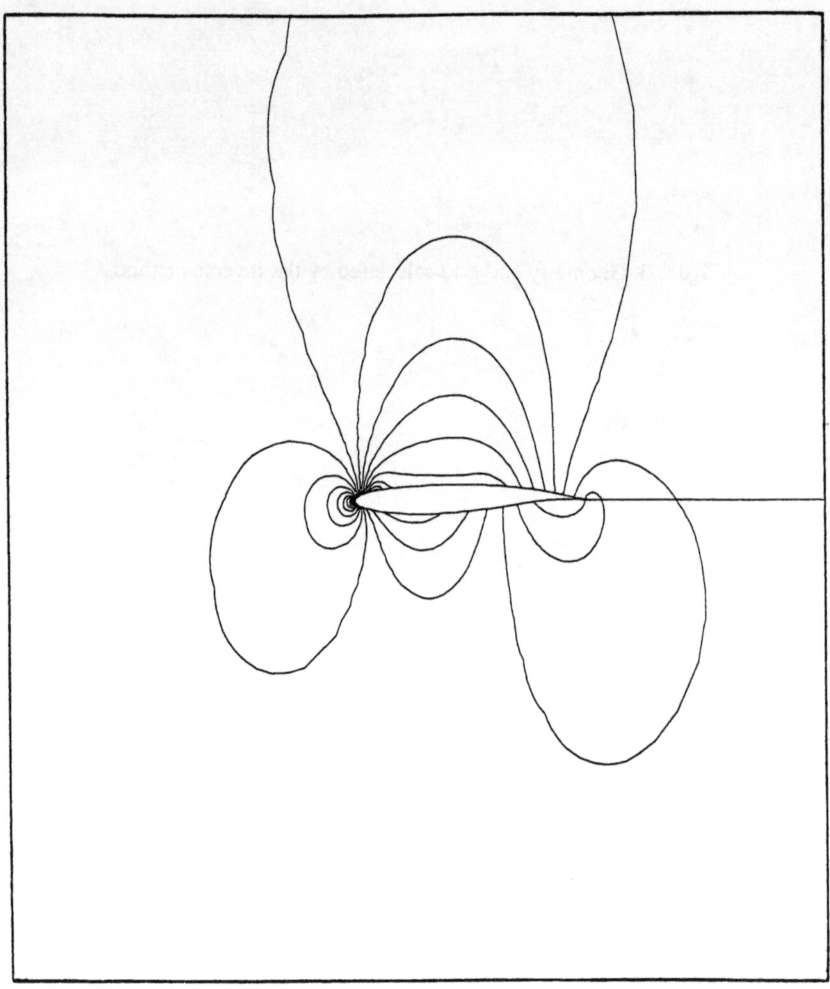

Figure 6. Mach contours of direct calculation for $M_\infty=.75$ and $\alpha=0^o$ ($M_{min}=.05$, $\Delta M=.05$).

A 165x21 C-type grid was generated on the (ϕ,ψ) plane. The computational cost associated with the inverse solution is of the order of 25 CPU secs on an Alliant FX 80 computer. The wall pressure coefficient distribution calculated by the direct method and used as boundary condition by the inverse method is shown in Fig.8. The difference between

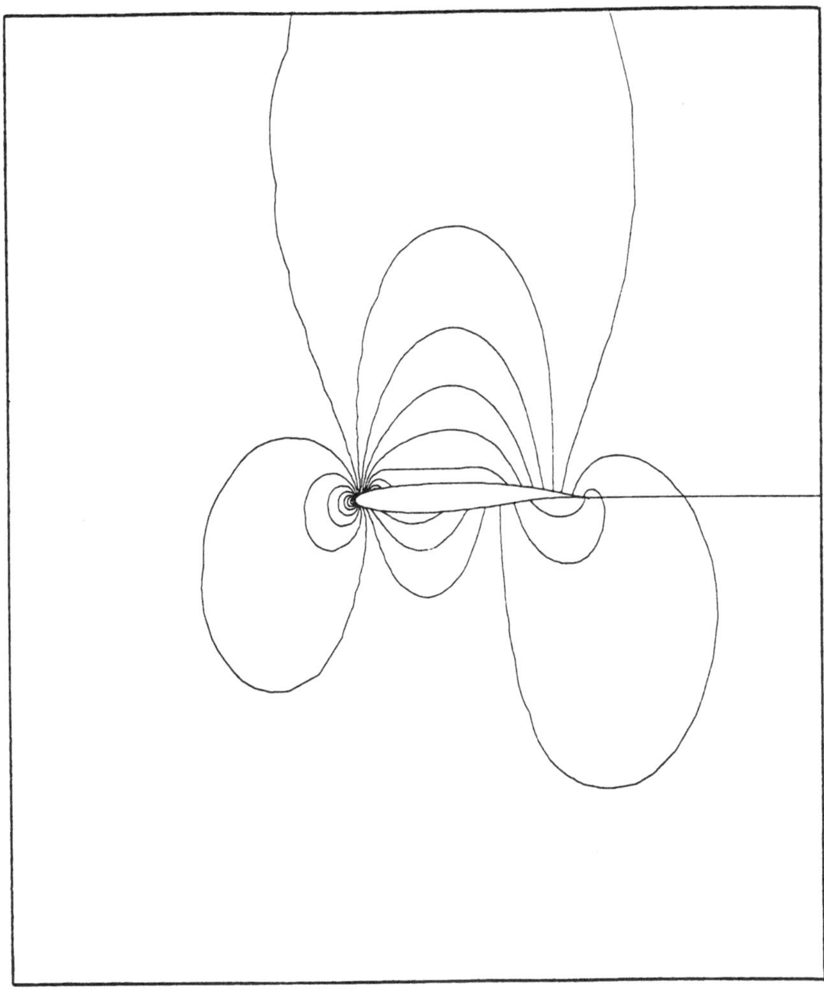

Figure 7. Isobar contours of direct calculation for $M_\infty=.75$ and $\alpha=0^o$ ($Cp_{min}=-.8$, $\Delta Cp=.1$).

computed airfoil coordinates and the corresponding target ones is presented in Fig.9. The l^2-norm distance between calculated and target profiles is of the order $5x10^{-4}$. This indicates that the accuracy of the reproduction procedure is very good, taking into account that some

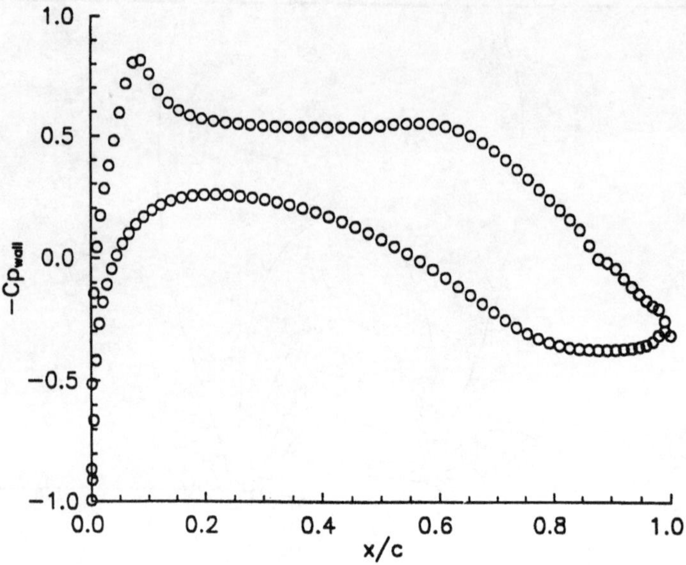

Figure 8. Wall pressure coefficient distribution calculated by the direct solver.

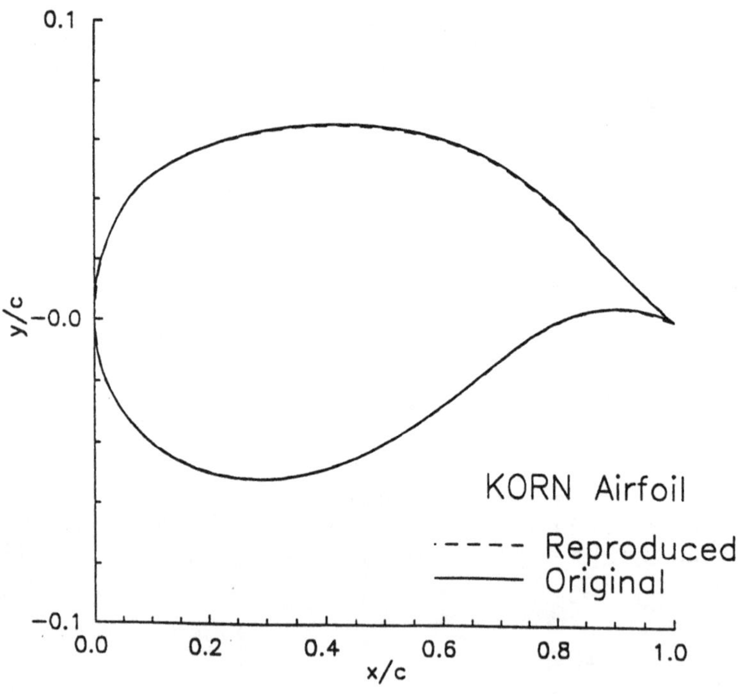

Figure 9. Comparison of original and reproduced Korn airfoil.

discrepancies are introduced by the interpolations and the different discretization schemes employed by the direct and inverse solvers.

T17 TEST CASE   3-D Subsonic Nozzle

Test case T17 concerns the reproduction of a 3-D subsonic, double turning converging nozzle. The geometry of the nozzle has been defined analytically. M_∞ at the inlet plane is set to 0.2 leading to a high subsonic exit Mach number of the order of .75.

A full-potential 3-D solver is employed in order to obtain the flow field and the velocity distribution on the lateral walls of the nozzle. The wall velocity distributions calculated by the direct solver are used as input by the 3-D inverse method in order to reproduce the geometry of the nozzle. It should be realized, however, that the 3-D inverse solver requires the velocity distributions along the boundary streamlines which, in general, do not coincide with the boundary grid lines of the direct solver. One has to translate, therefore, the flow field calculated by the direct solver to a form which can be comprehended by the inverse solver. For this purpose a post-processor has been coded, which performs the following tasks:

- Assuming that streamsurfaces (i.e. ψ=const., η=const. lines) are uniformly distributed at the inlet plane, a streamline computation is carried out. The procedure is based on the integration of the following transport equations

$$\vec{V}\cdot\nabla\psi = 0 \qquad , \qquad \vec{V}\cdot\nabla\eta = 0 \tag{21}$$

  where $\vec{V}$ is the velocity (vector) field calculated by the direct solver. Boundary conditions for ψ and η are required on the inlet plane only. The numerical integration of these two transport type equations is performed using an upwind approximate factorization scheme providing the values of the potential ϕ and streamfunctions ψ,η at the grid nodes of the original geometry.

- Respecting the limiting values of the ϕ,ψ,η variables a uniform rectangular computational grid was generated on the (ϕ,ψ,η) space.

- The velocity magnitude is then determined on the grid nodes of the computational (ϕ,ψ,η) space (which do not coincide with the original geometry grid nodes). A second order accurate interpolation procedure, using Taylor series expansion, is employed for this purpose.

Evidently, one expects that the numerical errors accumulated in this interpolation procedure will affect to some extend, the accuracy of the "reproduction". In order to minimize the numerical errors involved, the direct solver employs a 30x15x15 computational grid instead of the 30x7x7 one, which has been proposed originally.

A 43x15x15 uniform grid was generated on the (ϕ,ψ,η) space. The computational cost associated with the inverse problem solution is of the order of 650 CPU secs on an Alliant FX 80 computer. The projections of the center-lines of the original and the reproduced nozzle geometries on the (x,z) and (y,z) planes are presented in Figs 10a,b respectively. It should be noted that the center-line of the reproduced geometry is not a direct output of the inverse method. The geometry of the center-line of the reproduced nozzle is determined by averaging the coordinates of the four streamline vertices. This practice is acceptable since the nozzle has square cross sections. Three views of the reproduced nozzle geometry are presented in Fig.11. It is seen that the boundary nodes along the perimeter of the cross sections are not equidistant (as in the original geometry) since they represent the intersections of the cross sections with

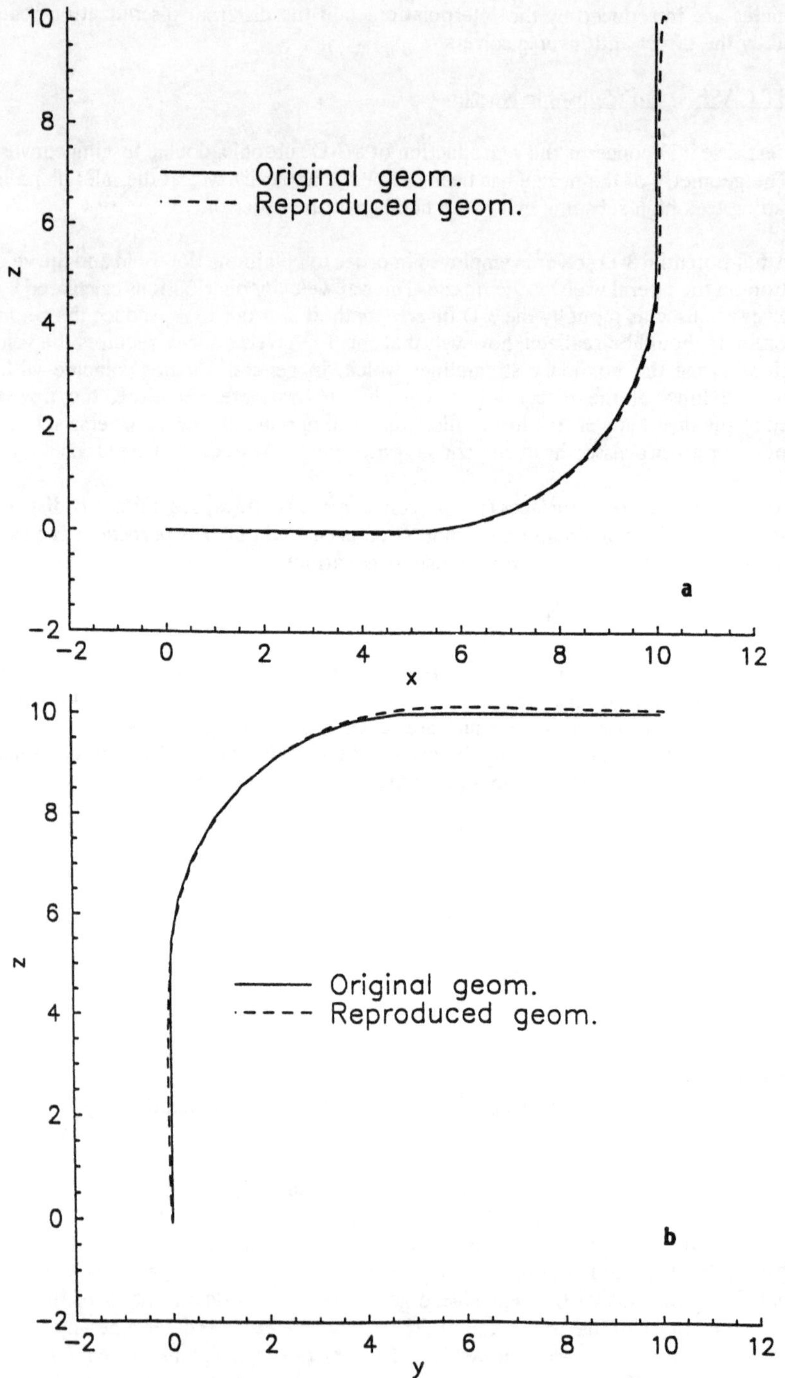

Figure 10. Projections of the original and reproduced geometry center-line on (a) the (x-z) plane and (b) the (y-z) plane.

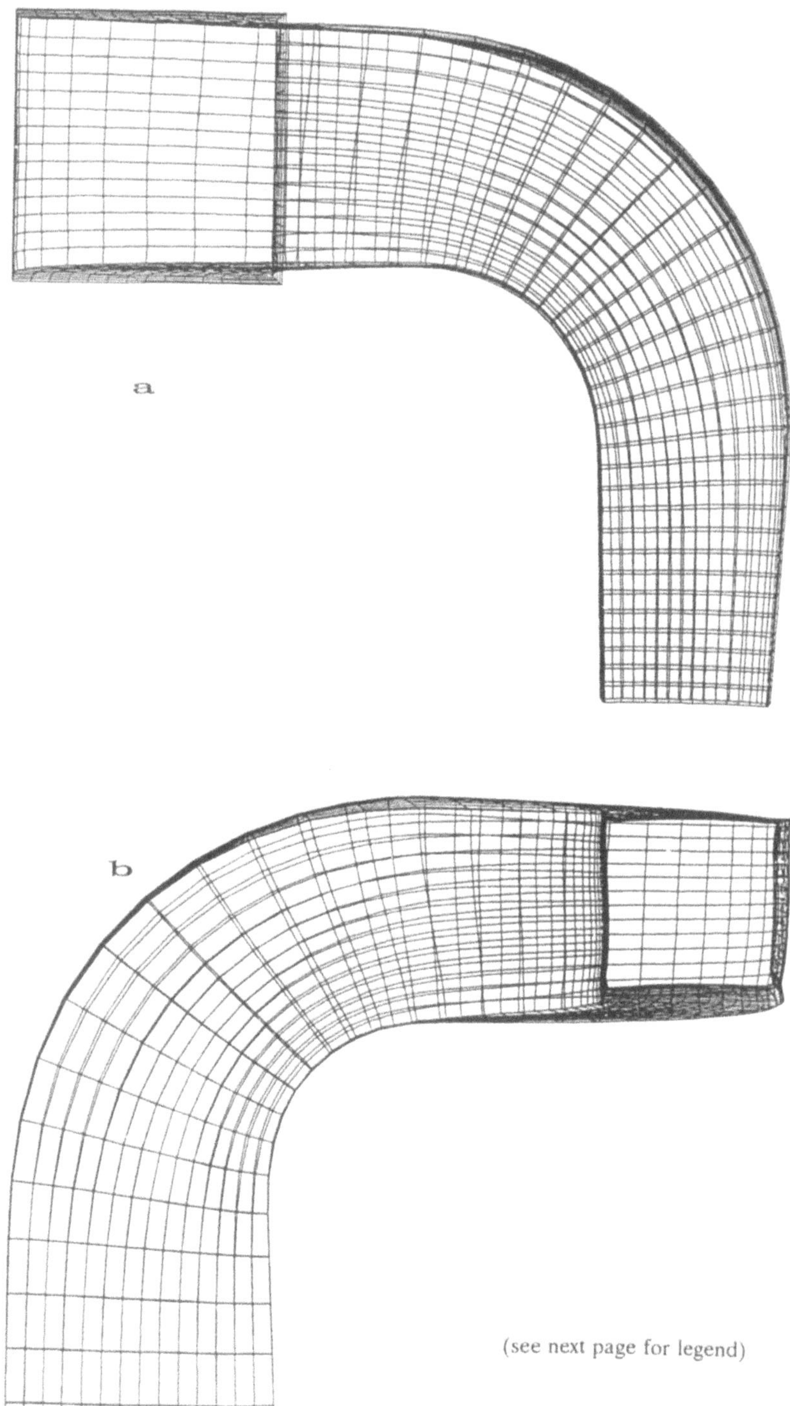

(see next page for legend)

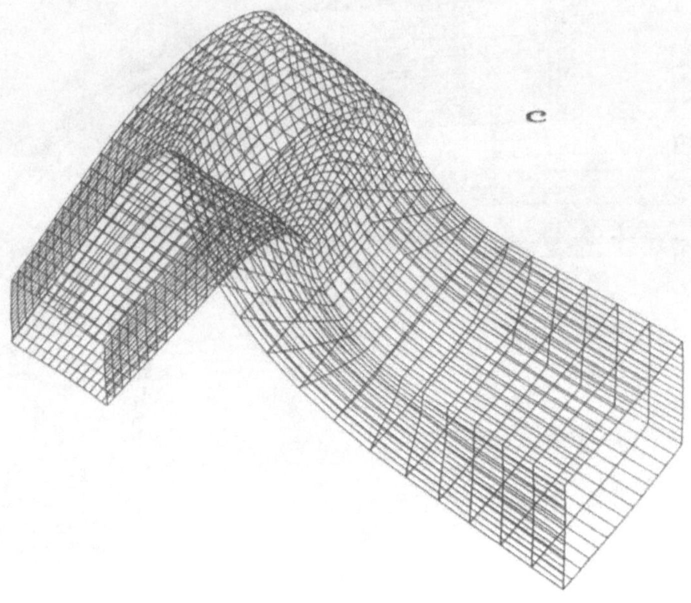

Figure 11. Views of the reproduced 3-D nozzle: (a) front view, (b) side view and (c) perspective view.

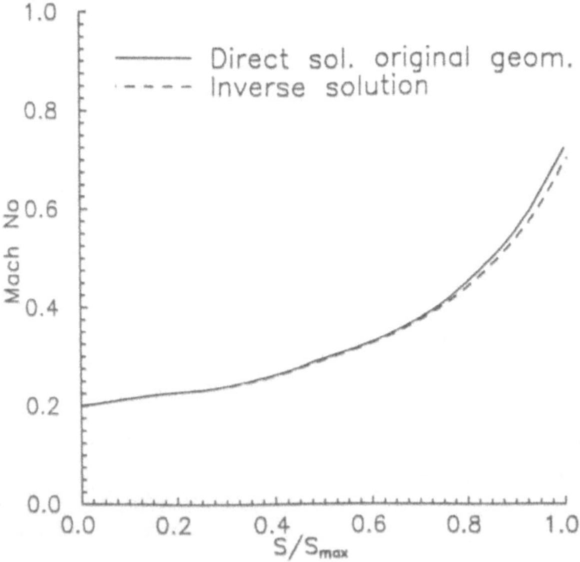

Figure 12. Mach number distributions of inverse (—) amd direct (---) calculations along the 3-D nozzle center-line.

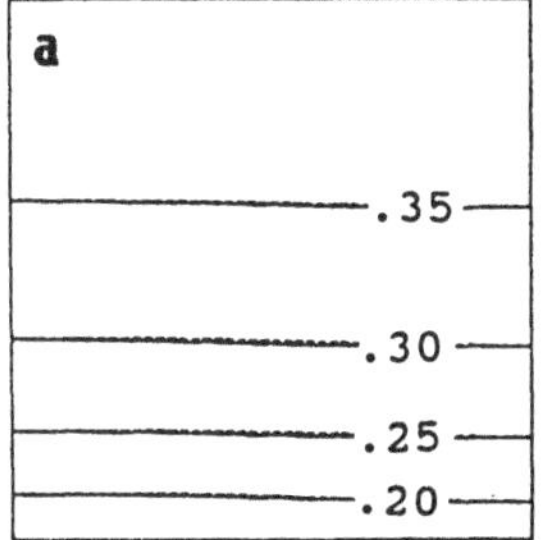

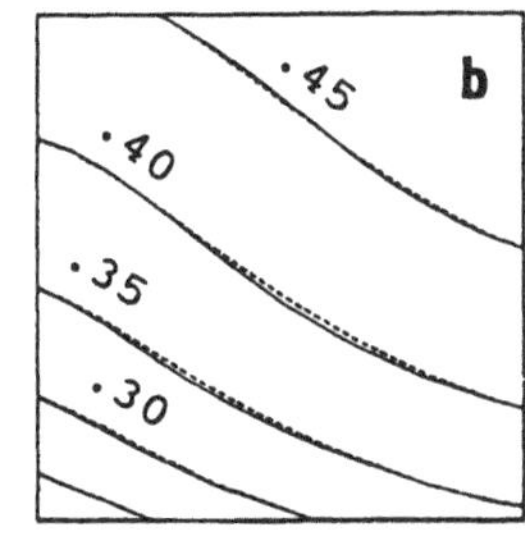

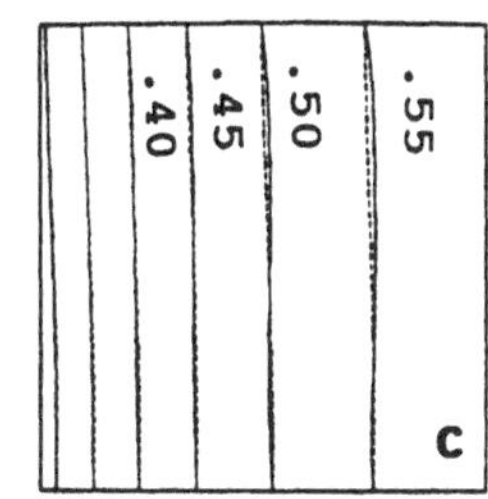

Figure 13. Mach number contours of inverse (—) amd direct (---) calculations on three sections normal to the centerline at (a) $s=1/3\ s_{max}$, (b) $s=1/2\ s_{max}$ and (c) $s=2/3\ s_{max}$

the boundary streamlines. The satisfactory comparisons of both the center-line and the lateral wall geometries indicate the accuracy of the proposed 3-D inverse method. Some discrepancies, however, are introduced by the interpolations and the different discretization schemes which are used in the direct and inverse solvers. Small discrepancies of the geometry near the nozzle vertices are due to the singular behaviour of the edge-streamlines.

Inverse and direct solvers Mach number distributions along the center-line of the nozzle are compared in Fig.12. Inverse and direct solvers Mach number contours on three cross sections normal to the center-line at the locations $s=1/3,1/2,2/3\ s_{max}$, s being the center-line arc length, are presented in Fig.13. In spite of the fact that a 3-D interpolation procedure was used to produce these contours the reproduction is quite accurate.

In all test cases, the required workshop output plots concerning the convergence history of the reproduction procedure are irrelevant, simply because the present design method is a single-pass one. In that sense, no initial guess for the geometry to be reproduced is required.

CONCLUSIONS

An inverse method for the solution of the potential compressible target pressure problem has been described. The method has been developed in terms of "natural coordinates". The potential function and two orthogonal streamfunctions (vectors) are used as independent variables, whilst the velocity magnitude and the aspect ratio of the elementary stream tubes cross section are considered to be the dependent ones. Differential geometry and generalized tensor analysis arguments are employed in order to formulate the method and derive the governing equations - an elliptic p.d.e. for the velocity magnitude and a second order o.d.e. for the streamtube aspect ratio. The present method is a single-pass, i.e. fast, one since the flow field which is governed by these two equations is provided without requiring any feedback from the geometry. The geometry is determined independently by integrating Frenet equation along the grid lines, after the flow field has been determined. The method is applicable to both 2-D and 3-D internal or external configurations, the 2-D case representing a particular

-reduced- 3-D case. In the framework of the present workshop, the method has been applied to the 2-D and 3-D nozzle test cases, T1 and T17 respectively, and to a 2-D airfoil test case, T4, with very satisfactory results.

REFERENCES

[1] LIGHTHILL, M.J.: "A new method of two-dimensional aerodynamic design", ARC R&M 2112 (1945).

[2] STANITZ, J.D.: "Design of two-dimensional channels with prescribed velocity distributions along the channel walls", NACA Report 1115 (1953).

[3] DULIKRAVICH, G.S.: "Aerodynamic shape design", AGARD-R-780 (1990), pp. 1-1 to 1-10.

[4] LABRUJERE, Th.E., SLOOFF, J.W.: "Computational methods for aerodynamic design of aircraft components", NLR TP 92072 U (1992) to appear in Annu. Rev. Fluid Mech., 25 (1993).

[5] STANITZ, J.D.: "General design method for three-dimensional potential flow fields. I-Theory", NASA CR 3288 (1980).

[6] STANITZ, J.D.: "General design method for three-dimensional potential flow fields. II-Computer Program DIN3D1 for simple unbranched ducts", NASA CR 3926 (1985).

[7] DEDOUSSIS, V., CHAVIAROPOULOS, P., PAPAILIOU, K.D.: "A 3-D inverse methodology applied to the design of axisymmetric ducts", ASME paper 92-GT-290 (1992).

[8] CHAVIAROPOULOS, P., DEDOUSSIS, V., PAPAILIOU, K.D.: "On the fully 3-D inverse potential target pressure problem. Part I: Method formulation and theoretical aspects" in preparation (1993).

[9] CHAVIAROPOULOS, P., DEDOUSSIS, V, PAPAILIOU, K.D.: "Compressible flow airfoil design using natural coordinates", to appear in Computer Methods in Appl. Mechanics and Eng. (1993).

[10] SYNGE, J.L., SCHILD, A.: "Tensor calculus", Dover Publ. INC., New York 1978.

[11] SAAD, Y., SCHULTZ, M.M.: "GMRES: A generalized minimal residual algorithm for solving nonsymmetric linear systems", SIAM J.Sci.Statist.Comput., 7 (1986) pp. 856-869.

[12] DEDOUSSIS, V., CHAVIAROPOULOS, P., PAPAILIOU, K.D.: "On the fully 3-D inverse potential target pressure problem. Part II: Numerical aspects and application to duct design" in preparation (1993).

[13] RIZZI, A.: "Computational mesh for transonic airfoils: The standard mesh", in Notes on Numerical Fluid Mechanics Vol. 3, Vieweg, Braunschweig (1981), pp. 222 to 253.

4. DESCRIPTION OF THE GRAPHIC SOFTWARE USED FOR THE WORKSHOP

The workshop workshop on optimum shape design in aeronautics took place in Barcelona at the Universitat Politècnica de Catalunya (UPC) on June 1992. Numerical results obtained by the participants on the different test cases were presented using the interactive graphic software FLAVIA developed at the International Center for Numerical Methods in Engineering (CIMNE). CIMNE is a research organization based in UPC and devoted to promote training and R & D activities in the field of numerical methods and their applications for solution of engineering problems.

FLAVIA (Flow Analysis VIsuAlizer) is a graphic software package based on the PHIGS standard specially developed for visualization of flow computations using finite element methods. More specifically, FLAVIA was developed at CIMNE during 1989-1992 in the framework of BRITE project P-2029 and since then it has been continuously updated.

FLAVIA can be effectively used for two and three dimensional flow visualization using standard and non structured finite element meshes. The basic capabilities of FLAVIA include visualization of: finite element meshes, velocity vectors, line and solid contours of Mach number, pressure and temperature fields, streamlines, particle tracking, changes of a specific flow parameter along a line, contours of flow parameters (Mach number, pressure, temperature, etc.) over a prescribed plane or surface (for 3D situations), etc.

FLAVIA is currently operational in workstations (Silicon Graphics, Hewlett-Packard, SUN, etc.) equipped with PHIGS standard capabilities.

Numerical results were provided by the workshop partners in tape form and automatically loaded in a Silicon Graphics workstation for interactive display during the workshop. Figures 1-7 show some of the results obtained by UPC for T1 and T4 test cases.

The successful experience proves that FLAVIA is an effective tool for interactive visualization of flow computations using finite element procedures. Extensions of FLAVIA accounting for flow results obtained from finite differences and finite volume methods are straightforward and they are currently in progress. FLAVIA is fully documented in English and accessible from CIMNE. For further information in FLAVIA please write to:

Dr. Gabriel Bugeda
International Center for Numerical
Methods in Engineering (CIMNE)
Universitat Politècnica de Catalunya (UPC)
C/ Gran Capità s/n
Campus Nord UPC; Mòdul C1
08034 Barcelona-Spain
Tel. ++34/3/401 64 94
Fax. ++34/3/401 65 17

LIST OF FIGURES

Workshop test case T1

Workshop test case T4

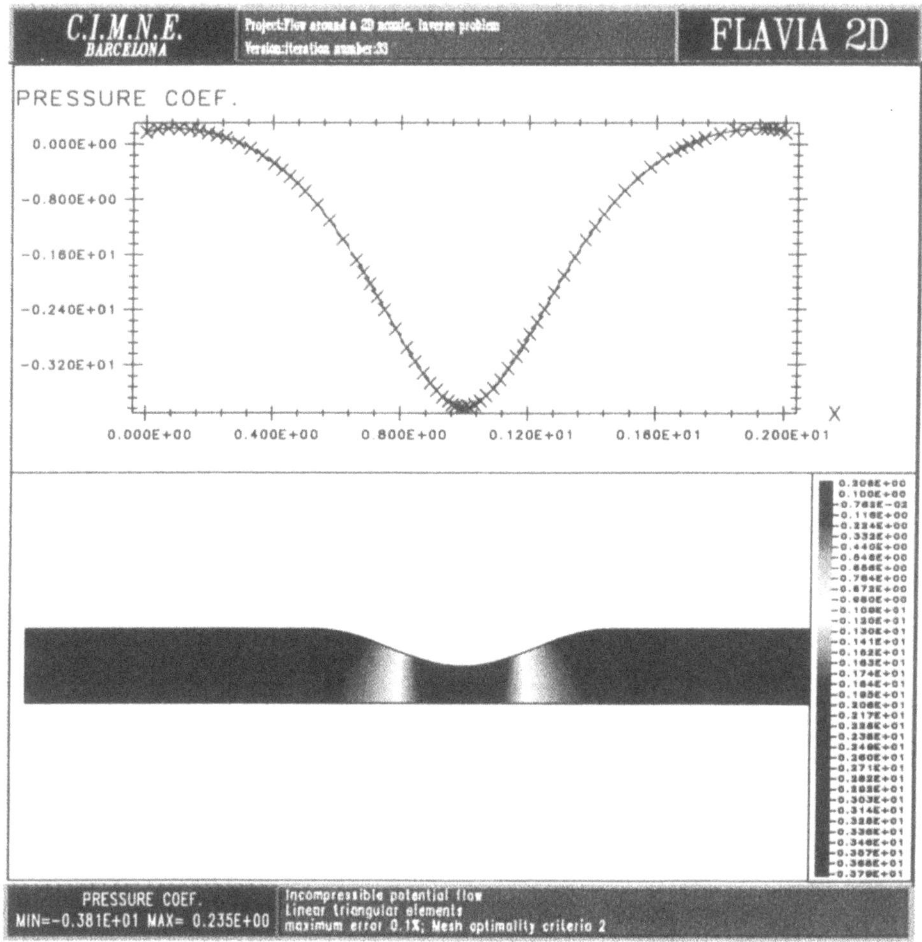

Figure 1. T1 test case: C_p contours on last design. C_p distribution over the nozzle and superposition of the target one.

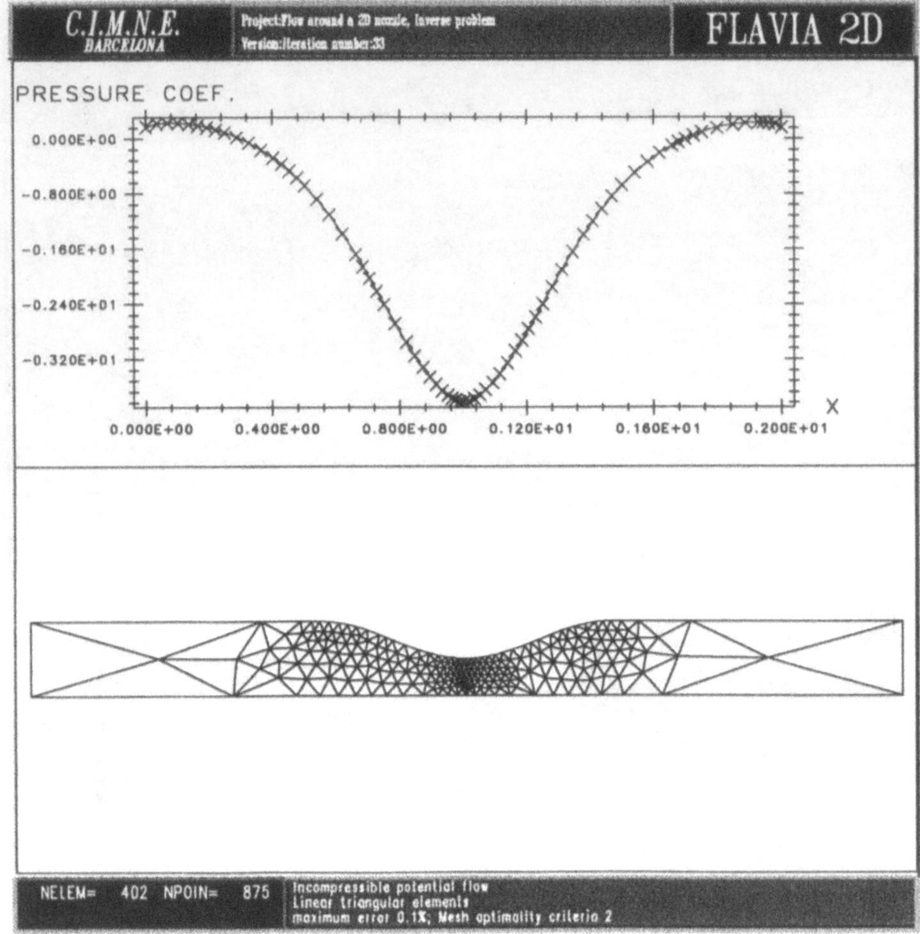

Figure 2. T1 test case: Finite element mesh corresponding to the last design. C_p distribution over the nozzle and superposition of the target one.

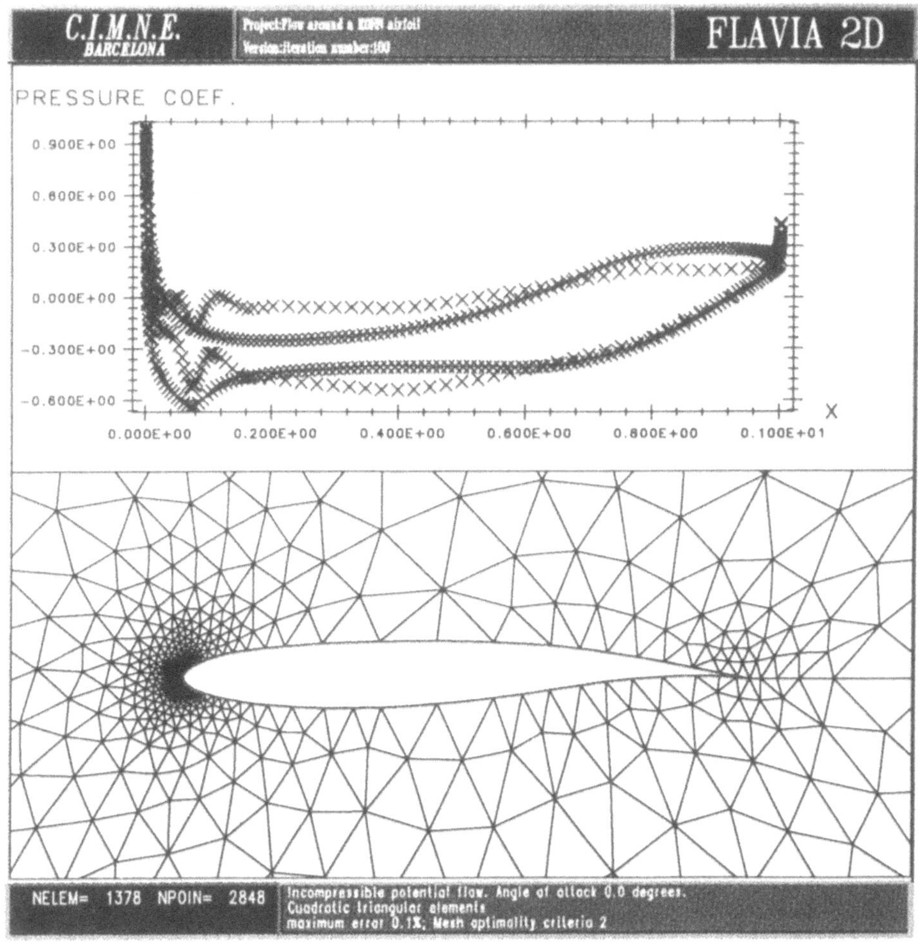

Figure 3. T4 test case: Finite element mesh corresponding to the last design. C_p distribution over the nozzle and superposition of the target and the initial ones.

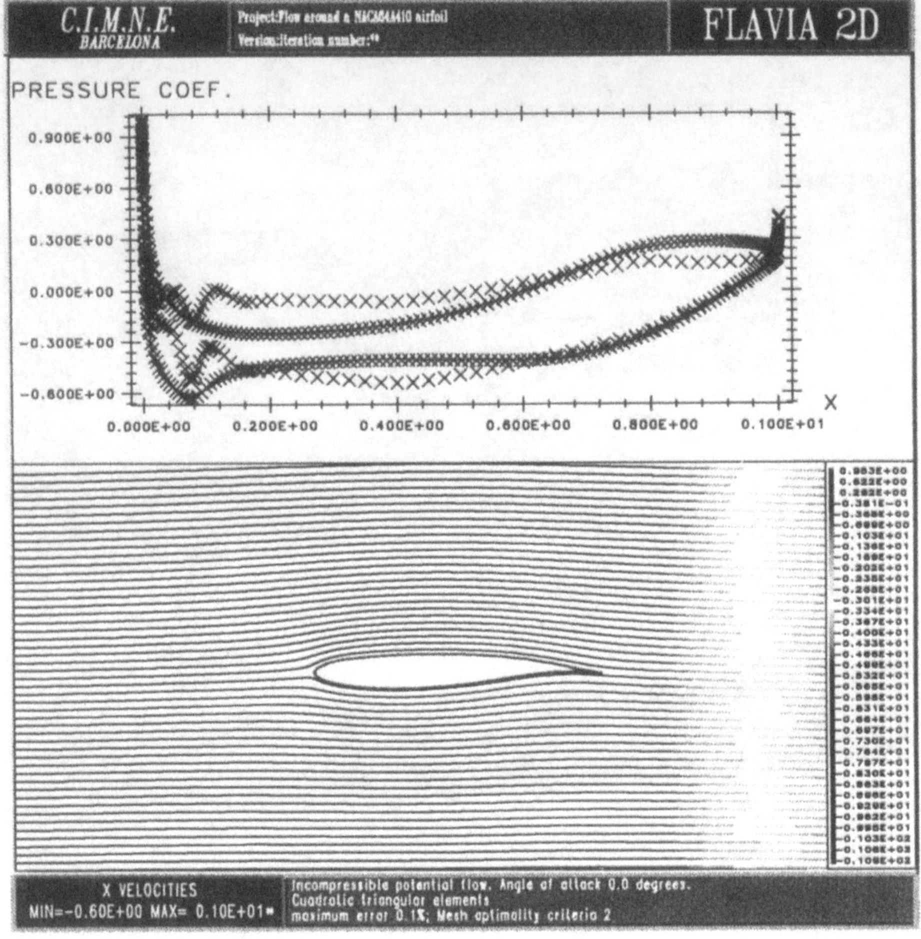

Figure 4. T4 test case: Particle tracking around the last design.

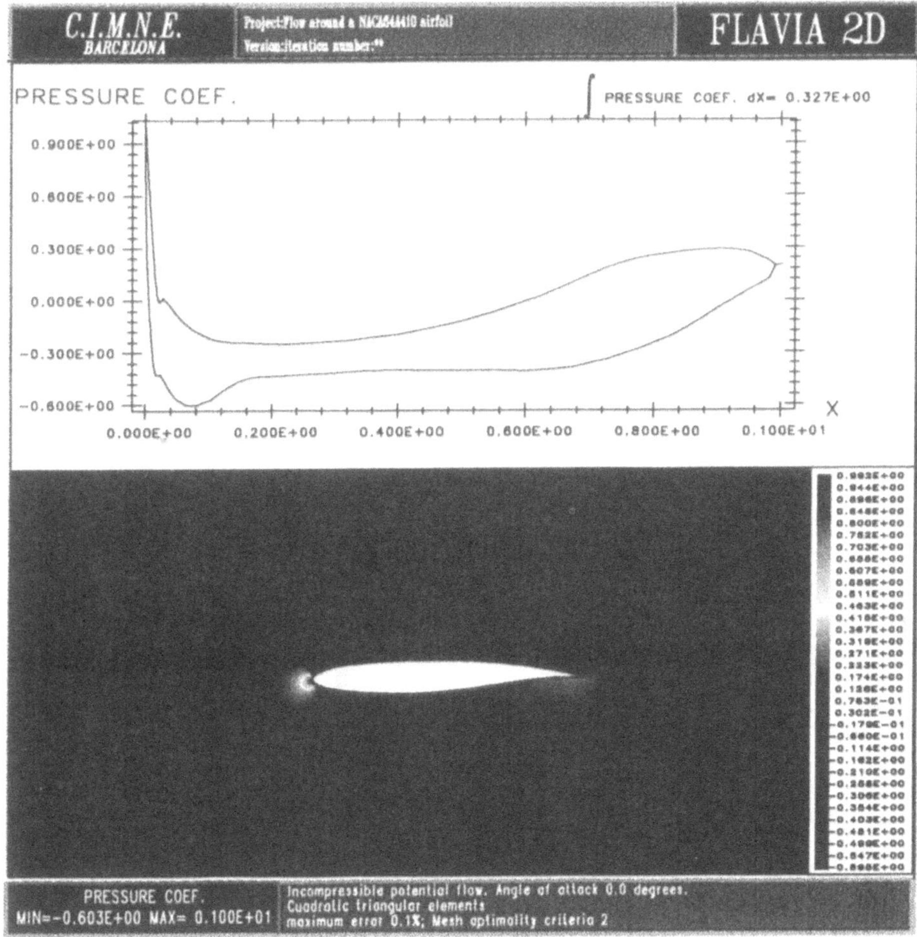

Figure 5. T4 test case: C_p contours and distribution around the last design.

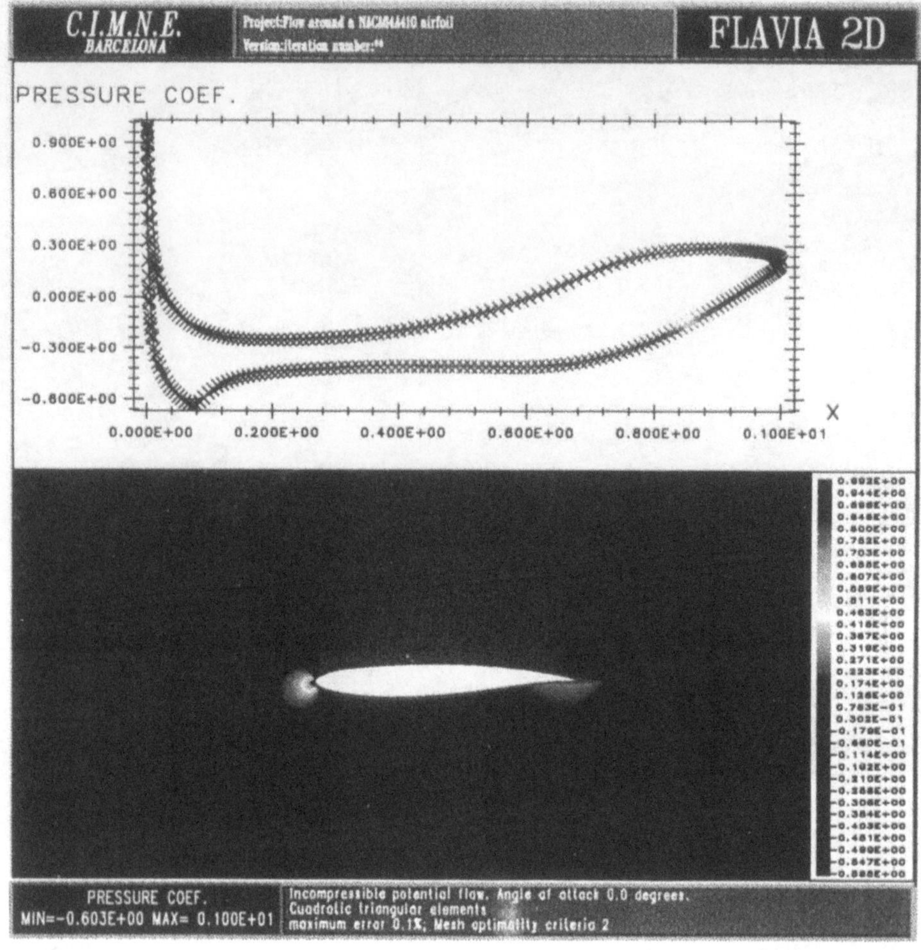

Figure 6. T4 test case: C_p contours and distribution around the last design and superposition of the target one.

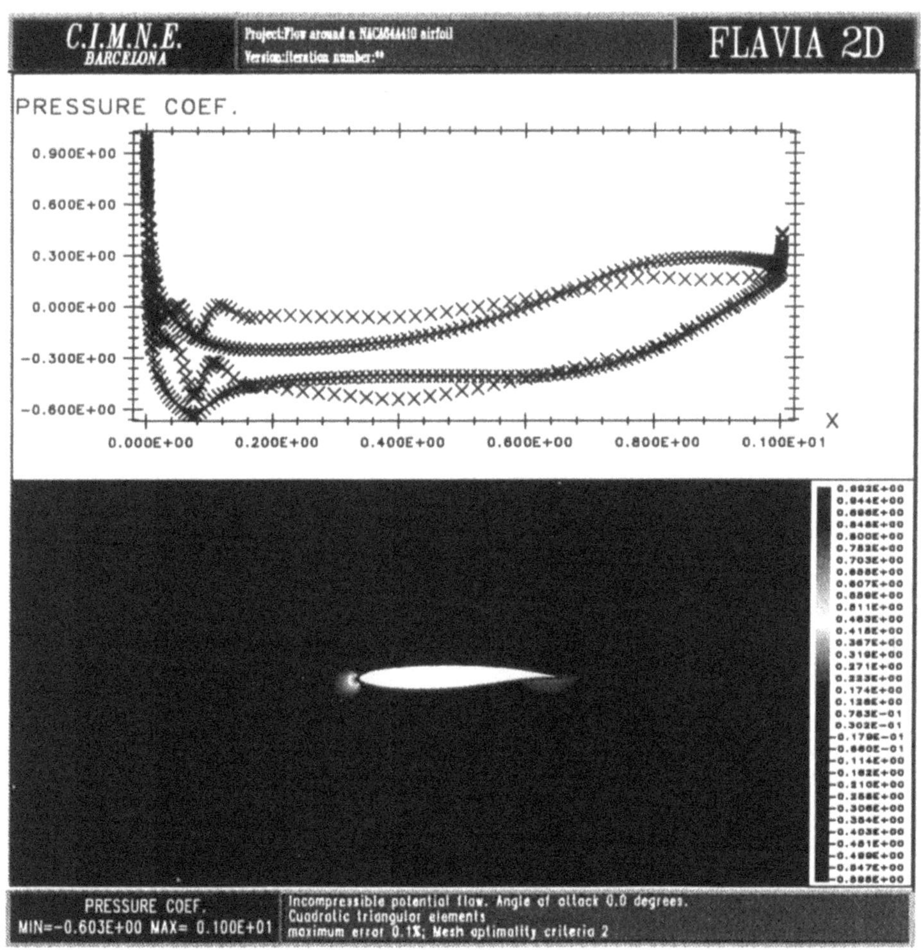

Figure 7. T4 test case: C_p contours and distribution around the last design and superposition of the initial and target ones.

5. SYNTHESIS OF THE WORKSHOP TEST CASES

This section contains the synthesis of the different contributions presented for the resolution of the workshop test cases. This synthesis contains, for each test case, the reasons that make the test case interesting for the workshop, the special difficulties for the resolution, its possible interest for a future workshop in the context of the ECARP Brite/Euram project and a comparison of the results presented by different contributors.

Table 1 contains some general information concerning the contributions presented for each test case. It shows the number and the name of the contributors for each test case and some information about the tools used for the solution. This table shows that only a limited number of test cases have been performed by at least three partners (test cases T1, T4 and T6) making comparisons of results a rather difficult task. Only one partner (NLR) run multi point test cases with viscous corrections which still require large and expensive computer facilities.

Different approaches (choice of approximation and optimizers) have been used by contributors as shown in Tables 2 and 3.

The computed results have been obtained with design softwares of partners for the assesment of methodologies implemented duting the project. A detailed of description of each numerical optimization tools can be found in the corresponding author's contributions presented in Chapter 4.

The methodologies and contributions of each partner are the following:

1. DASSAULT AVIATION
   - *methodology:* investigation of inverse and optimization problems with finite elements using unsteady transonic flow analysis solver; these problems are solved by optimal control theory with gradient methods or GMRES exact derivative of the cost function.
   - *contribution to the workshop:* inverse and optimization test problems T4 and T6 related to airfoil shape design in transonic flows.
2. DEUTSCHE AIRBUS
   - *methodology:* 3D panel code with mixed "design" and "analysis" patches parametrized by Bezier splines embedded in an inverse procedure which minimize the sum of square pressure deviations to trat combination of multi wing/body/pylon componenet aircraft systems.
   - *contribution to the workshop:* definition, preparation of input data, design specification and results of test cases T4, T10 and T14.
3. ALENIA
   - *methodology:* optimization code for minimization of drag with target lift coefficients, reconstruction test cases and constrained multi point transonic design.
   - *contribution to the workshop:* optimization test case T6 and reconstruction test case T4.
4. NLR
   - *methodology:* inverse algorithm for "flexible multi point wing design" targeting specific pressure distributions and geometric constraints at different design points with a full potential flow analysis solver.

Table 1: Contributions to the workshop test cases

test case	partner	control variables	flow variables	computer	CPU	conver.
T1	INRIA	33	4(Euler)* 423*31	Convex C210	15.320 s	8 ord.
	UPC	7	Variable (betw. 107 and 940)	Personal Iris 35 TG	1.080 s	5 ord.
	NTUA	61	61*31	Alliant FX-80	8 s	One shot. L2 norm: $2 * 10^{-4}$
T2	INRIA	6	4(Euler)* 423*31	Convex C210	1.700 s	0,78 ord.
	UPC	6	Variable (betw. 107 and 1531)	Personal Iris 35 TG	600 s	0,25 ord.
T3	INRIA	33	4(Euler)* 423*31	Convex C210	4.000 s	2,75 ord.
T4	DASSAULT	46	1.103	IBM820	600 s	2 ord.
	DEUTSCHE AIRBUS	8*2				3 ord.
	NLR	94	94			
	INRIA	20		IBM820	200 s	3 ord.
	NTUA		165*21	Alliant FX80	25 s	
	UPC	25	Variable $\sim$ 2.700	Indigo R4000	3,5 h	3 ord.
T6	ALENIA					2 ord.
	INRIA		2.700			
	DASSAULT	20	3.108			
T8	NLR	94	10.000	CDC Cyber 962	700 s	0.5 ord.
T10	DEUTSCHE AIRBUS	62	126	Apollo WS 30		3 ord.
T14 recons.	DEUTSCHE AIRBUS	826	65	Apollo WS 30	6-8 h	4 ord.
T14 design	NLR	1.900	140.000	NEC SX3	2.600 s	1,5 ord.
T17	NTUA	4*30*15	30*15*15	Alliant FX80	650 s	One shot.

- *contribution to the workshop:* definition and computation of the 2D multi point design test case T8 and also the single point 3D wing design test case T14; computation of the reconstruction test case T4.

5.1 INRIA ROCQUENCOURT

- *methodology:* airfoil shape optimization with finite elements in transonic potential lifting flows using GMRES or exact differentiation of the criteria.
- *contribution to the workshop:* reconstruction of the transonic shockless Korn airfoil test case T4 and the drag optimization RAE2822 for transonic conditions.

Table 2: Approximations of the flow analysis solvers

	DASSAULT	DEUTSCHE AIRBUS	ALENIA	NLR	INRIA	NTUA	UPC
Panels		X		X			
Finite Differences			X			X	
Finite Volumes		X		X	X		
Finite Elements	X				X		X

Table 3: Optimization or inverse method

	DASSAULT	DEUTSCHE AIRBUS	ALENIA	NLR	INRIA	NTUA	UPC
One pass method						X	
Hierarchical					X		
Steepest Descent					X		
Residual Correction				X			
Levenberg Marquardt		X					
Conjugate Gradient	X				X		
Feasible Direction			X				
BFGS			X				X
Non Linear GMRES	X				X		

5.2 INRIA SOPHIA ANTIPOLIS

- *methodology:* hierarchical method relying on a multilevel approach for the academic problem of shape optimization of a nozzle in a subsonic or transonic Euler flow; capability of solving with a "one shot method" approach mesh variables, flow equation and shape optimality conditions simultaneously.
- *contribution to the workshop:* shape reconstruction and optimization of an Euler flow in the nozzle geometry defined by test cases T1, T2 and T3.

6. NTUA

- *methodology:* 3D inverse method theory for potential flows and its application on complicated fully 3D test cases.
- *contribution to the workshop:* definition and computations of 2D and 3D nozzle flows T1, T2 and T17 for reconstruction and optimization problems; computation of test case T4 for the reconstruction of the transonic shockless Korn airfoil problem.

7. UPC
- *methodology:* control of the error involved in the resolution of the flow model equations for each design, error estimator to control the quality of the results and the use of first and second order exact sensitivity analysis of all the magnitudes involved in the design process. Implementation of the methodology with P2 triangular finite elements using a potential flow analysis solver.
- *contribution to the workshop:* reconstruction and optimization of nozzle geometries in the T1, T2 and T14 test cases with a finite element potential code using adaptive remeshing.

The following pages contain more specific information for each test case.

TEST CASE T1

Choice of the test case

Test case T1 addresses the reproduction of a simple symmetric nozzle under subsonic flow conditions. Because of its simplicity the case is ideal for a first evaluation of the developed design tools. The original geometry (area of interest $0 \leq x \leq 2$) was extended upstream and downstream for inlet and exit for uniformity reasons. This extension caused local discontinuities of the second derivative of the nozzle boundary at the junction points. Test case 1 allows for a straight-forward selection of the design parameters (y-coordinates for fixed x-coordinates of the boundary points) and regriding procedure (if needed).

Difficulties of the Design with the test case

The only difficulty which can be attributed to this test case is the lack of C2 continuity of the extended boundary which may lead to local inaccuracies when the flow solver employs high order discretization schemes.

Possible Interest for the ECARP project

The subsonic nozzle reproduction test case with the viscous effects included may be an ideal candidate for the future ECARP workshop. Although the geometry is rather simple the corresponding viscous flow field has interesting features because of the adverse pressure gradients induced by the divergent part of the nozzle. If a realistic Reynolds number is selected the throat-inlet area ratio should be relaxed in order to prevent separation in the decelerating part of the flow.

Brief synthesis of contributions

Test case 1 was considered by the groups of INRIA, UPC and NTUA. INRIA used a first order accurate upwind Euler solver while UPC and NTUA used full potential solvers for the cost function evaluation. Applying the hierarchical parametrization method INRIA code reduced the cost function 5 order of magnitude (compared to its initial value) in 100 approximately cost function evaluations while the UPC code had the same performance in 30 iterations using, however, 7 (compared to the 33 of INRIA) design parameters. A similar analogy holds for the convergence of the gradient norm. The combination of the exact gradient computation with the hierarchical parametrization renders the INRIA method and efficient one when a large number of design parameters is considered. The NTUA method used 33 design variables and the final result needed the equivalent effort of one cost-function evaluation (being a single-pass approach). The L2 norm distance of the target and solution geometries is converged 2.3 (INRIA) and 3.0 (UPC) orders, while the corresponding distance of the NTUA solution is 2.1×10^{-4}. For the last case it appears that the maximum error occurs in the neighbour of the C2 discontinuity (curvature discontinuity). An interesting feature of the UPC and NTUA methods is that a high quality adapted grid is automatically generated during the optimization procedure.

TEST CASE T2

Choice of the test case

The optimization of a simple symmetric nozzle geometry under subsonic flow conditions is considered. The inlet cross section and the inlet to throat area ratio of the nozzle are fixed, acting as geometrical constraints in the optimization process. The cost function is defined as a global measure of the pressure slope along the horizontal axis of symmetry. Similar to test case 1, the geometry is extended upstream and downstream the area of interest, for inlet and exit flow uniformity reasons. The case was selected because of its simplicity concerning the geometrical aspects of the problem (selection of the design parameters, regriding procedure) but, also, because it addresses several real life applications. Typical applications involve the design of wind-tunnel components, pipes and junctions, afterbodies, turbomachinery diffusors, etc. The main objective in nozzles design is the control of the boundary layers in terms of viscous drag control and blockage factor reduction (in particular at lower Reynolds numbers $0(10^5)$). The local (and global) pressure slope along the wall is the key point for the boundary layer control.

Difficulties of the Design with the test case

The present test case, despite its simplicity, proved that the definition of cost functions for flow optimization is not a trivial task. Numerical problems occured and non-admissible shapes were designed when the cost pressure integral was considered along the axis of symmetry instead of the wall itself (which consists a more physical approach). The power coefficient of the pressure integral plays an important role, as well, affecting the smoothness of the optimized pressure distribution and, consequently, the geometry solution. Using the original definition of the cost-function the participants were obliged to introduce a small number of design parameters (compared to test case 1) in order to prevent numerical instabilities and oscillations formation on the optimized shape. When the wall pressure integral was considered the above problems were totally suppressed.

Possible Interest for the ECARP project

Test case 2, suitably redefined, will be an interesting optimization problem for the ECARP workshop. The geometrical constraints may be retained but the inlet to throat area ratio should be relaxed (if the length is fixed) in order to prevent flow separation in the deceleration part of the flow. The cost function, defined along the solid wall, could be the total drag, including the pressure drag and the skin friction integrals. The Mach number should be restricted in the subsonic domain, in order to prevent shock-boundary layer interaction problems.

Brief synthesis of contributions

Test case 2 was considered by the groups of INRIA and UPC. INRIA used a first order accurate upwind Euler solver, while UPC used a full potential solver. Triangular meshes were used for spatial discretization by both participants. Due to its non-symmetrical properties and low accuracy discretization scheme (inviscid fluxes) the Euler solver produces non-symmetric solutions in symmetric geometries and vice-versa. In that respect, the symmetry if the shape was explicitly imposed during the iterations. Applying the hierarchical parametrization strategy INRIA code reduced the cost function two orders of magnitude in 70 iterations approximately, using 6 parameters only. The same number of parameters was used by UPC without, however, obtaining a converged solution with the initial definition of the cost function. The UPC method, employing P2 spatial discretization schemes (compared to the P1 of INRIA) appears to be more sensitive to the power coefficient of the pressure slope. Although the encountered numerical instabilities are not fully understood it is believed that, at least part of them, they are driven by the incomplete definition of the cost function along the axis of symmetry (the extension parts are excluded) in combination with the geometrical constraints and the smoothness properties of the cost function. Analyzing the results of UPC it is clear that the sudden change of the pressure slope in the junctions and throat regions is reflected on the boundary geometry which obtains a wavy shape.

TEST CASE T3

Choice of the test case

Test case T3 concerns the optimization of a transonic nozzle. The definition of the geometry, the constraints and the cost function is identical to that of test case 2. The inlet Mach number is changed, however, from 0.2 to 0.5. This change is expected to cause sonic conditions at the throat section leading to the formation of a strong shock wave (at least for the starting geometry) in the rear part of the nozzle. The actual location of the shock wave depends on the back pressure value. According to the 1-D theory, two isentropic (shock-free) solutions exist, one for subsonic and a second for supersonic exit conditions. If the back pressure is equal to the inlet pressure (as in our case) and the geometry is symmetric in respect to the y-axis, the aft-throat flow should follow the subsonic isentropic branch minimizing, thus, the pressure slope integral since no shock (pressure discontinuity) is any more present. In that respect and according to the 1-D theory, the optimized nozzle shape should not differ much from that of case

2. In practice, however, serious discrepances from the 1-D theory are expected because of the two-dimensionality of the flow. An asymmetric (in the x-sense) solution is more probable which will smear the shock along the horizontal axis of symmetry.

Difficulties of the Design with the test case

The difficulties encountered in test case 2 are present here, as well, amplified from the transonic nature of the flow. However, since INRIA is the only participant in this test case no general conclusions may be stated. The INRIA solution presents certain irregularities, the most obvious of them being the extremely low maximum Mach number value obtained in the final solution. Actually, the complete flow field appears to be low subsonic, conflicting with the specifications of the problem. This is probably due to the low accuracy of the numerical scheme which was employed for the integration of the state equations. In that respect it is not clear whether the asymmetric geometry solution reflects the physics or it simply demonstrates the accumulative effect of the numerical errors.

Possible Interest for the ECARP project

Test case 3 is not suggested for the ECARP workshop. It is believed that the inclusion if viscous effects will create difficulties even for the convergence of the direct problem because of the strong shock/boundary layer interaction effect. Since the inviscid version of the test case is not yet clarified, a further increase if its complexity is not advisable.

TEST CASE T4

Choice of the test case

As it was addressed in the Introduction about the objectives of this Workshop, it appears of primary importance to check the capabilities of the different methodologies to treat inverse or optimization problems, firstly on a simple reconstruction problem. A two-dimensional inviscid flow test case has been chosen to evaluate the performance of the solvers.

This test case has been settled in that sense and has been issued from [1]. Target pressure distribution corresponds to the Korn airfoil taken at a shock-free design for a free stream Mach number of $M_\infty = 0.75$, and an angle of attack $\alpha = 0^{\underline{o}}$. Initial profile is taken as the NACA64A410 airfoil. At these conditions. a shock is located above the profile. The aim of this test case is to recover from this distribution with shock, the shock-free Korn airfoil.

Difficulties of the Design with the test case

In terms of specifically optimum shape design, this test case is nothing but a first (and necessary) stage in verifying and comparing the capabilities of the proposed methods. The main difficulty associated with this problem lies in the capture of a shock-free solution for the Korn airfoil. Some contributors show the presence of a small shock; this can be due to either the flow solver or a some inaccuracy in the shape representation by a too coarse approximation. This case gives opportunities to check the accuracy of shape representation modules.

Possible Interest for the ECARP project

Viscous effects have already been incorporated in a full potential solver by one of the participants (NLR). The pressure distribution which is compared to the inviscid one is somewhat different and the number of elementary problems needed to recover the Korn shape is twice the one of the inviscid case. This comparison between inviscid and viscous cases can provide trends about the extra cost of optimum shape design problems when considering viscous effects.

Brief synthesis of contributions

Table 4 summarizes the different flow solvers used by the participants in test case T4. In Table 4 the figures of the different computations performed by the participants are summarized. The meaning of the different terms in this table is the following:

NFLOW: denotes the number of discretization nodes of the flow solver.

NDOF: denotes the number of control parameters used for the optimization.

ITER: denotes the number of iterations (i.e. elementary flow problems) performed for the computation.

RES: denotes the number of iterations needed to reach the decrease of the residual given in parenthesis.

Table 4: Flow solvers used for test case T4

participants	flow solvers	optimization methods
DASSAULT	full potential finite element method	GMRES on functional
DEUTSCHE AIRBUS	full potential in body fitted coordinates	inverse Hessian combined with steepest descent
NLR	full potential panel method	equivalent incompressible perturbation velocity and residual correction
INRIA	full potential finite element method	GMRES on optimality conditions
NTUA	Velocity/Streamtube formulation with associated grid	direct inverse method
UPC	Incompressible potential finite element method	sensitivity analysis

The estimated CPU times needed to achieve the residual given in Table 5 are respectively 600' on scalar IBM820 (about 15 Mflops) for Dassault, 200' on scalar IBM820 for INRIA, 25" on Alliant FX80 for NTUA, 3.5 hours on Silicon Graphics Indigo R4000 for UPC.

Table 5: Comparisons for the reconstruction problem T4

participants	control parameters	NFLOW	NDOF	ITER	RES
DASSAULT	ordinates of mesh points	1103	46	600	600 (2)
DEUTSCHE AIRBUS	Bezier limits degree 10		8*2	75	75 (2)
NLR	panel	94	94	25	10 (3)
INRIA	Spline ordinates of control points		20	250	200 (3)
NTUA	Inverse method	165*21			
UPC	B-splines ordinates of control points	2700	25	50	50 (2)

TEST CASE T6

Choice of the test case

One point design

A drag minimization problem issued from [1] has been proposed to test the methods in a pure drag reduction problem. The chosen initial profile is the RAE2822 airfoil whose definition has been specified in the definition of the test case.

The freestream Mach number is taken to be M=0.730 and the initial angle of attack is $\alpha = 2^{\circ}$. The flow is supposed to be inviscid and modeled by either full potential or Euler equations. Two minimization problems are proposed.

1) Improvement of existing profile via an inverse problem

Firstly, we formulate the problem as a perturbation of an inverse problem. The target pressure distribution p_{tar} is taken to be the actual pressure distribution of the initial profile predicted by the numerical solution of the flow equations. The inclusion of this target distribution will force the method to generate a profile with a lift coefficient close to that of the initial profile. Drag coefficient C_D is added to the cost function I so that its expression becomes:

$$I = I_{tar} + \beta_1 C_D$$

where I_{tar} denotes the cost function corresponding to the pure inverse problem. The addition of a drag penalty now causes the method to reshape the profile to reduce its drag where β_1 is a parameter. The sensitivity of the minimization procedure and of the solution to the magnitude of this parameter can be studied.

2) Improvement of existing profile via a minimization

Secondly, we consider the previous problem as a pure minimization problem allowing the angle of attack to float. Two different ways to perform this problem can be investigated. The first one is to use a penalty method by minimizing an augmented objective function without constraints expressed as:

$$Minimize \;\; I = C_D + \beta_2 (C_L - C_{L_{tar}})^2 \tag{2a}$$

A second formulation is a minimization problem under constraints which can be expressed as:

$$\text{Minimize } I = C_D \text{ under the constraint of a given lift } C_{L_{tar}} \tag{2b}$$

Difficulties of the Design with the test case

This case is really an optimization problem aiming to reshape a profile to reduce its drag. It is concerned with the reduction of transonic drag generated by the induced shock at the upper surface.

Table 6: Flow solvers used for test case T6

participants	flow solvers	optimization methods
ALENIA	full potential finite diff. with conformal mapping	quasi-Newton like (BFGS) (no constraints) feasible direction (constraints)
DASSAULT	full potential finite element method	conjugate gradient
INRIA	full potential finite element method	GMRES on optimality conditions

Table 7: Comparisons for the drag reduction test case T6

participants	control parameteres	NFLOW	NDOF	ITER	RES
ALENIA 2a) $\beta = 1$ 2b)	Bezier curve segment				 200 (2) 140
DASSAULT 1)	Spline ordinates of control points	3108	20	—	—
INRIA 1)	Spline ordinates of control points	2700			

Table 8: Comparisons of obtained drag reduction

participants	case	LIFT	DRAG
ALENIA	2a) $\beta_2 = 1$ 2a) $\beta_2 = 100$ 2b)	1.046 1.047 1.046	0.0026 0.0076 0.0026
INRIA	1) $\beta_1 = 10$ 1) $\beta_1 = 50$		
Reference [1]	2b)	1.037	0.0016

Possible Interest for the ECARP project

Viscous effects have to be included in order to take into account possible separation of the viscous boundary layer linked to adverse pressure gradient. Drag reduction in this case will concurrently aim to reduce the shock strength and the onset of separation.

Brief synthesis of contributions

Tables 6, 7 and 8 show a comparison between the flow solvers, the optimization methods and the results of the different partners involved in test case T6.

TEST CASE T8

Choice of the test case

Test case T8 has been chosen because of its relationship with problems occurring at the design of transport aircraft outer wing sections (no flap, no slat) and at the design of helicopter rotor blade sections. The test case aims at the design of an airfoil which combines a favourable high speed and a favourable low speed performance. The test case concerns a two-point airfoil design with design criteria formulated as desired pressure distributions for two completely different operating conditions. The first target pressure distribution has been chosen to produce a favourable low speed high lift capacity (M=0.2 $\alpha = 7.8^{\underline{o}}$). The second target pressure distribution has been chosen for its favourable high speed performance (weak shock) (M=0.77 $\alpha = 0^{\underline{o}}$). Both pressure distributions are realistic in the sense that for each target an airfoil can be designed such that the pressure distribution will be approximated at the given onset flow condition. The pressure distributions have been obtained by applying a pressure distribution optimization code for viscous flow [2]. Hence, the case can serve as a test for any method considering multi-point design for subsonic/transonic viscous flow. And of course, it could also be used for single-point design in either subsonic viscous or transonic viscous flow. So, it can serve several purposes in the ECARP "Optimum Design" workshop.

Difficulties of the Design with the test case

The main problem with the multi-point test case is the fact that the target pressure distributions are most likely incompatible in the sense that it will not be possible to determine one airfoil shape which will produce these targets at the given or even at modified onset flow conditions. So, design calculations will lead to some kind of compromise at best. In general, it may be expected that by giving the different targets a different weight one of these targets will be approximated better than the other, as appears from the NLR results. Otherwise, the NLR results do not yet show how to deal with incompatibility in the targets. More research will be needed to solve this problem.

complete flow analysis. The comparison of the calculated wing section shapes with the target wing section shapes shows the limitations of using only a few sections for control of the wing geometry, but presumably it is only a matter of using more variables in order to get a closer approximation to the target.

TEST CASE T10

Choice of the test case

This test case deals with the reconstruction of a two-elements configuration from their pressure distribution. Each of the two elements is free to rotate around its leading edge. The objective of this test case is to check the performance of the reconstruction algorithms when they are applied to multi-element airfoils.

This case correspond to many realistic situations and, in particular, with a landing configuration where a flap profile is extended in addition to the main wing.

Difficulties of the Design with the test case

The main difficulties of the resolution of this test case arise from the mixture of two types of variables. This mixture is due to the two following aspects:

- A different set of design variables must be defined for each profile. It means that it is necessary to deal with two different set of design variables.
- It is necessary to fit not only the shape of each profile, but also the rotation angle of the profile with respect to the horizontal direction. From the dimensional point of view two different types of variables are needed, one of them with length units, and the other one with angle units.

Possible Interest for the ECARP project

This test case is still valid for use in the ECARP workshop though it would probably be more attractive with the inclusion of viscous effects.

Brief synthesis of contributions

There is only one contribution to this test case made by DEUTSCHE AIRBUS. The results of this contribution are very good and there are only small discrepances between the target and the obtained solution due to the different flow models used in boths cases.

TEST CASE T14

Choice of the test case

Test case T14 is related to transport aircraft wing design. Using the well known F4 configuration of which the body geometry and the wing geometry are specified by means of a number of contour points, two test cases have been formulated. For the purpose of checking the correctness of design algorithms, a so-called reconstruction test case has been defined. In such a test case, first the pressure distribution for a given configuration is calculated for given operational conditions. Then, starting with an arbitrary wing shape as initial guess, and with the calculated pressure distributions as design criteria, its shape must be reconstructed. For the purpose of demonstrating the applicability of design algorithms, a more realistic design test case has been defined.

Difficulties of the Design with the test case

Reconstruction

The reconstruction test case concerns an initial configuration which includes the F4 body, the F4 platform and the F4 wing root section but which has a perturbed wing geometry in comparison with the F4 configuration (perturbation in terms of camber and thickness). Part of the geometry of the F4 configuration has to be reconstructed according to the following steps:

- Calculation of the pressure distributions on the F4 wing-body configuration for $\alpha = 6^{\underline{o}}$ and $M = 0.3$.
- Application of the design algorithm with the wing pressure distribution from the first step as target and with the initial configuration as starting point.

By changing the onset flow conditions the test case can be used perfectly well for design methods which have been developed for other flow regimes than considered in the present application and as such can be used directly in the ECARP "Optimum Design" workshop.

The test case should not present particular unforeseeable difficulties because, being a partial reconstruction case, it is known beforehand that a solution to the problem exists. Nevertheless, the design method to be applied should be capable of determining completely new shapes of the wing sections.

Design

The objective of the more realistic design exercise is to improve the (inviscid) low speed high lift pressure distribution for a known wing-body configuration by modifying sing section contours, while retaining the body shape, the platform of the wing and the wing root section. In this case, the starting point is the F4 configuration itself. The target pressure distribution is specified by means of chordwise pressure distributions in a number of wing sections. This target pressure distribution has been obtained by firstly calculating the pressure distribution on the wing by means of the NLR potential flow solver [3], [4] for $C_L \approx 1,60$ for the original wing-body configuration at $M = 0.3$. Secondly, this pressure distribution has been modified to reduce the high supersonic velocity peaks to subsonic values, while retaining approximately the sectional lift coefficients; the lower surface pressure distribution has not been changed.

This test case tests a design method in particular at a point where most methods known from the literature are weakest, namely the determination of the wing section nose shapes.

Possible Interest for the ECARP project

This test case is still valid for use in ECARP "Optimum Design" workshop though it would be probably more attractive to move to transonic flow conditions. The latter will imply, however, the respecification of the target pressure distributions.

Brief synthesis of contributions

The NLR results for the design test case confirm the expectation that the largest geometry modifications concern the nose shapes. Nevertheless, it appears possible to arrive at a rather close approximation of the desired pressure distribution, albeit that "man-in-the-loop" activities were necessary. The results demonstrate clearly the usefulness of the residual correction approach. Unfortunately, comparison with other methods has not been possible because NLR appeared to be the only workshop participant with successful computations for this test case.

The wing design results shown by Deutsche Airbus concern the reconstruction test case. For that case, they demonstrate the effectiveness of the "equivalent normal velocity" concept used for calculating the gradient information as required in the minimization process. The presentation of the convergence history by L2 as a function of the "number of elementary problems" (aero code calls) seems to be somewhat misleading with respect to computing costs involved. According to their method description a call to the aero code during the determination of the Jacobian involves only a fraction of the computations needed for a complete flow analysis. The comparison of the calculated wing section shapes with the target wing section shapes shows the limitations of using only a few sections for control of the wing geometry, but presumably it is only a matter of using more variables in order to get a closer approximation to the target.

TEST CASE T17

Choice of the test case

The reproduction of a 3-D duct is considered in this test case. The duct geometry is composed by two converging elbows turning in perpendicular directions. The flow is subsonic with a wide range of Mach number variations from 0.4 to 0.8 approximately. The target pressure distribution is provided by a direct full potential code. The corresponding flow field is fully three-dimensional and this fact motivated the definition of the test case. Several real-life applications, such as turbomachinery and aircraft air-inlets, are related to 3-D ducts design. In such cases, the objective of the duct shape optimization involve the boundary layers control (directly related to the pressure distribution on the walls) and, additionally, the control of the flow distortion at the exit plane. The definition of the design parameters is not a straightforward task because of the lack if a privileged direction in which the problem could be considered as quasi-two dimensional (like the 3-D wing case).

Difficulties of the Design with the test case

Due to the above mentioned practical difficulties the case was considered by the NTUA group only. Using an "one shot" inverse method this group was able to reproduce the original shape with admissible accuracy and with relatively low computational effort.

Possible Interest for the ECARP project

Due to the reduced number of contributions in the EUROPT 1 phase this test case is not suggested for further investigation in the ECARP workshop.

REFERENCES

[1] Jameson, A. *Aerodynamic optimum shape design*, AIAA paper 92-0670, AIAA Aerospace Sciences Meeting, Reno (USA), Jan. 1992.

[2] van Egmond, J. A. *Numerical optimization of target pressure distributions for subsonic and transonic airfoil design* NLR TP 89155 L.

[3] van der Vooren, J., van der Wees, A. J. and Meelker, J. H. *Transonic potential flow calculations about transport aircraft* AGARD Conf. Proc. No. 412 (1986).

[4] van der Vooren, J., van der Wees *Inviscid drag prediction for transonic transport wings using a full-potential method* AIAA-90-0576 (1990).

6. CONCLUSIONS AND FURTHER COMMENTS

The most significant outcomes of the workshop on "Optimum Design in Aerodynamics" concern the progress accomplished in the following methodologies:

- Design with Euler solvers.
- Fast one pass inverse methods for rotational flows.
- Hierarchical multi level methods for control variables.
- automatic adaptive remeshing.

Among these methodologies several softwares experimented by the contributors to the workshop seem to be quite promising for industrial applications. The most promising developments are the following:

1) Inverse variant of a 3D panel code capable to deal with complete wing-pylon-nacelle-fuselage configurations (Deutsche Airbus).
2) Exact gradient optimizers with Euler and Navier Stokes solvers via automatic symbolic differentiation requiring reasonable programming efforts (INRIA and Dassault Aviation).
3) Hierarchical methods providing higher efficiency when combined with optimization and parametrized methods (INRIA).
4) Use of fast "one shot" methods with optimality conditions (INRIA).
5) Extension of the inviscid potential-stream function method introduced by Stanitz to rotational flows (NTUA).
6) Residual correction method for single and multi-point transonic inverse design (NLR).

The comparison of the results shows that the discrepancy of the computed solutions for optimization and inverse problems is much larger than for direct ones solved by flow analysis solvers.

Some of the difficulties encountered in the above design methods can be explained by the following remarks:

Remark 1: Existence and/or uniqueness of the solution for optimization problems is still an open problem, specially for transonic multi point design problems.

Remark 2: Classical numerical optimization tools can trap design solutions into local minima depending on the initial guess. Other search space methods like genetic algorithms could provide local minima.

Remark 3: The number of control variables play a major role in the definition of the optimization problems. Bezier splines appear to be the best parametrization for accuracy.

Remark 4: The computational effort can be reduced with the use of hierarchical parametrization.

The results presented in the workshop by the different contributors suggest some new directions of research:

i) Improvement of the accuracy of the design by 3D automated adaptive remeshing and appropriate choice of control variables.

ii) Acceleration of the convergence by one shot methods and hierarchical parametrization.

iii) Systematic use of new parallelizable design algorithms on new MIMD parallel architectures.

It can be noticed that the best computed designs during the workshop were obtained by optimization softwares containing part of the ingredients mentioned in the above items.

The optimization of designs with complex viscous flows on industrial geometries remain still a challenge.

It appeared that for practical 3D applications including viscous effects a major effort remains still to be done (a target in ECARP optimum design project). In that direction, the achievement of cost effective and accurate designs with available novel methodologies will undoubtedly require HPCN tools on parallel computers.

Addresses of the Editors of the Series "Notes on Numerical Fluid Mechanics"

Prof. Dr. Ernst Heinrich Hirschel (General Editor)
Herzog-Heinrich-Weg 6
D-85604 Zorneding
Federal Republic of Germany

Prof. Dr. Kozo Fujii
High-Speed Aerodynamics Div.
The ISAS
Yoshinodai 3-1-1, Sagamihara
Kanagawa 229
Japan

Prof. Dr. Bram van Leer
Department of Aerospace Engineering
The University of Michigan
3025 FXB Building
1320 Beal Avenue
Ann Arbor, Michigan 48109-2118
USA

Prof. Dr. Michael A. Leschziner
UMIST-Department of Mechanical Engineering
P.O. Box 88
Manchester M60 1QD

Prof. Dr. Maurizio Pandolfi
Dipartimento di Ingegneria Aeronautica e Spaziale
Politecnico di Torino
Corso Duca Degli Abruzzi, 24
I-10129 Torino
Italy

Prof. Dr. Arthur Rizzi
Royal Institute of Technology
Aeronautical Engineering
Dept. of Vehicle Engineering
S-10044 Stockholm
Sweden

Dr. Bernard Roux
Institut de Mécanique des Fluides
Laboratoire Associé au C.R.N.S. LA 03
1, Rue Honnorat
F-13003 Marseille
France

Brief Instruction for Authors

Manuscripts should have well over 100 pages. As they will be reproduced photomechanically they should be produced with utmost care according to the guidelines, which will be supplied on request. In print, the size will be reduced linearly to approximately 75 per cent. Figures and diagrams should be lettered accordingly so as to produce letters not smaller than 2 mm in print. The same is valid for handwritten formulae. Manuscripts (in English) or proposals should be sent to the general editor, Prof. Dr. E. H. Hirschel, Herzog-Heinrich-Weg 6, D-85604 Zorneding.

Notes on Numerical Fluid Mechanics (NNFM) Volume 55

Volume 35 Proceedings of the Ninth GAMM-Conference on Numerical Methods in Fluid Mechanics (J. B. Vos / A. Rizzi / I. L. Ryhming, Eds.)

Volume 34 Numerical Solutions of the Euler Equations for Steady Flow Problems (A. Eberle / A. Rizzi / E. H. Hirschel)

Volume 33 Numerical Techniques for Boundary Element Methods (W. Hackbusch, Ed.)

Volume 32 Adaptive Finite Element Solution Algorithm for the Euler Equations (R. A. Shapiro)

Volume 31 Parallel Algorithms for Partial Differential Equations (W. Hackbusch, Ed.)

Volume 30 Numerical Treatment of the Navier-Stokes Equations (W. Hackbusch / R. Rannacher, Eds.)

SPRINGER NATURE

GPSR Compliance

The European Union's (EU) General Product Safety Regulation (GPSR) is a set of rules that requires consumer products to be safe and our obligations to ensure this.

If you have any concerns about our products, you can contact us on ProductSafety@springernature.com

In case Publisher is established outside the EU, the EU authorized representative is:

Springer Nature Customer Service Center GmbH
Europaplatz 3
69115 Heidelberg, Germany

Zeitfracht Medien GmbH
Ferdinand-Jühlke-Straße 7
99095 Erfurt, Deutschland
produktsicherheit@kolibri360.de